T0367784

CURRENT DEVELOPMENTS IN MATHEMATICAL BIOLOGY

Proceedings of the Conference on
Mathematical Biology and Dynamical Systems

Series on Knots and Everything — Vol. 38

CURRENT DEVELOPMENTS IN MATHEMATICAL BIOLOGY

Proceedings of the Conference on Mathematical Biology and Dynamical Systems

The University of Texas at Tyler 7 – 9 October 2005

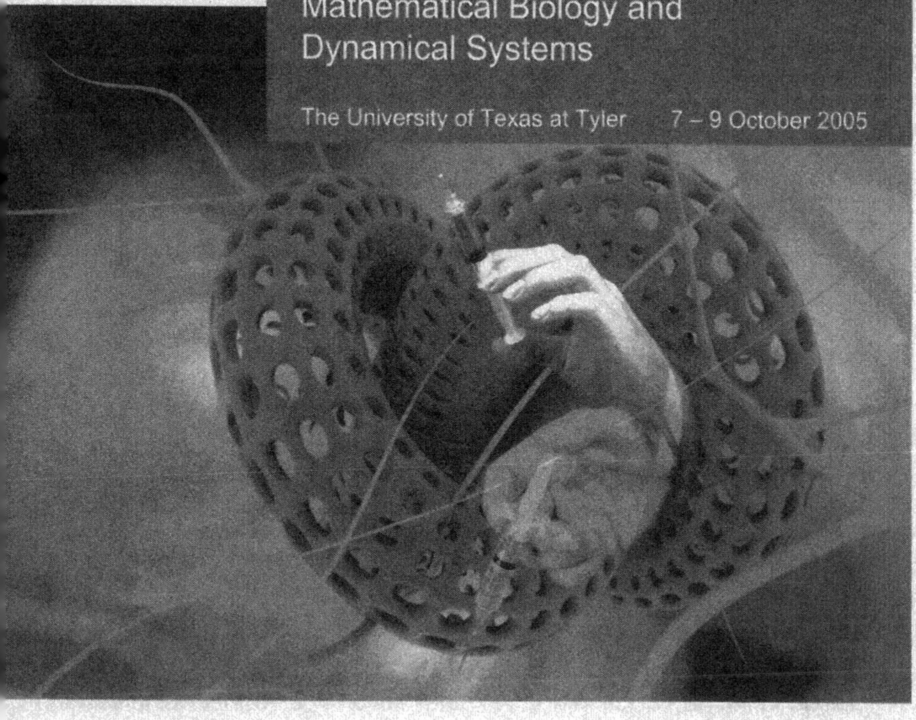

Editors

Kazem Mahdavi, Rebecca Culshaw & John Boucher *The University of Texas at Tyler, USA*

World Scientific

NEW JERSEY • LONDON • SINGAPORE • BEIJING • SHANGHAI • HONG KONG • TAIPEI • CHENNAI

Published by

World Scientific Publishing Co. Pte. Ltd.
5 Toh Tuck Link, Singapore 596224
USA office: 27 Warren Street, Suite 401-402, Hackensack, NJ 07601
UK office: 57 Shelton Street, Covent Garden, London WC2H 9HE

British Library Cataloguing-in-Publication Data
A catalogue record for this book is available from the British Library.

Series on Knots and Everything — Vol. 38
CURRENT DEVELOPMENTS IN MATHEMATICAL BIOLOGY
Proceedings of the Conference on Mathematical Biology and Dynamical Systems

ISBN-13 978-981-270-015-5
ISBN-10 981-270-015-3

Typeset by Stallion Press
Email: enquiries@stallionpress.com

Printed in Singapore by World Scientific Printers (S) Pte Ltd

INTRODUCTION

This volume is the Proceedings of the Conference on Mathematical Biology and Dynamical Systems that was held at The University of Texas at Tyler, October 7–9, 2005.

The aim of this conference was to bring together a group of established and up-and-coming researchers from a diverse range of topics in Mathematical, Computational and Theoretical Biology, to share their latest results in a setting that would encourage informal discussion of future research directions. The conference also provided an opportunity for young researchers and graduate students to interact with top researchers and to be exposed to the very latest areas of research.

There were twelve one-hour plenary talks during the two and a half days that the conference ran.

The subjects covered truly reflected the wide spectrum of Mathematical and Theoretical Biology, and researchers came from campuses from all over the United States, from the University of Florida to Cornell University in New York, to The Utah State University. Talks covered subjects as wide-ranging as predator-prey models in the chemostat, using DNA as a computational tool, immunology and T-cell dynamics, and evolutionary dynamics.

This proceedings contains eight papers by eight different authors and covers articles with applications to immunology, ecology, computation and more.

The Conference was generously supported by the National Security Agency and The University of Texas at Tyler. The University of Texas at Tyler made an ideal setting for this conference, and we are especially grateful to the library staff and faculty in the Mathematics and Biology departments for the work they did in making this conference a success.

CONTENTS

METHYLATION OF DNA MAY BE USEFUL AS A COMPUTATIONAL TOOL: EXPERIMENTAL EVIDENCE

SUSANNAH GAL, NANCY MONTEITH, SARA SHKALIM,
HU HUANG and TOM HEAD

Department of Biological Sciences and
Department of Mathematics,
Binghamton University, Binghamton,
NY 13902-6000, USA

August 13, 2006

Previously we have explained the abstract concept we call 'aqueous computing' and illustrated it with concrete wet lab results. Here, we explore the use of methylase enzymes to 'write' on double-stranded DNA molecules at sites where restriction enzymes will cut if, and only if, the sites have not previously been methylated. A site represents the bit zero (False, F) if the site has been methylated and the bit one (True, T) if it has not been methylated. 'Reading' is done by attempting a cut at each of the sites. We found 8 commercially available methylases and 8 corresponding restriction enzymes that would not cut after the action of one of the methylases. We were able to confirm that methylation by each of these 8 enzymes individually blocked cleavage only by the restriction enzyme associated with that site and not any other enzyme. We then used these enzymes to approach a 3-variable, 4-clause satisfiability (SAT) problem using either plasmid DNA (pBluescript) or PCR product made from the region containing the restriction enzyme sites on the plasmid. Pairs of methylases were defined to represent each of the states of the operators p, q and r, one methylase for p and another for p', etc. We methylated the DNA in parallel at the two sites so either the p site was methylated (making p false) or the p' site was methylated (making p' false). We did that for the other two variables as well to create a set of logically consistent DNA fragments. Then we applied the 4 clauses using restriction enzymes to cut DNA fragments that did not satisfy them. At the end, we found evidence for intact DNA indicating an answer satisfying all of the clauses. To confirm the state of each of the Boolean operators, we used cleavage by the appropriate restriction enzyme. We found in the computation with both the plasmid and the PCR product, one site pair to show false in both sites; q and q', for instance. This should not be possible. We suspected incomplete cutting during the clauses by one of these restriction enzymes, specifically BssHII. In summary, we did successfully show the usefulness of DNA methylation in a scheme to do a mathematical computation. Thus, we have added to our arsenal of potential methods of performing DNA computing in the aqueous style.

1. Introduction

An increasing number of groups in the world are exploring the possibilities of using DNA as a computer register. This has at least one obvious advantage over traditional computers, namely there are 4 possible bits at each position – 'A', 'C', 'G', and 'T'. Also, there are many naturally occurring enzymes that work with DNA that can be harnessed to facilitate DNA computing. Another advantage of using DNA in aqueous solution (i.e., in water) is that with a simple split-and-mix technique, problems that require an exponential number of steps to solve conventionally, involve only a linear number of steps with this approach. Previously we have explained this abstract concept which we call 'aqueous computing' and illustrated it with concrete wet lab results (for reviews see [3, 6]). Each aqueous computation begins with a vast number of molecules all of identical structure. Each of these molecules is used as a 'nano-tablet' having prescribed locations at which bits can be 'written' by specified means. This writing step requires an alteration such that bits where "writing" has occurred can be distinguished experimentally from "non-written" areas. We have generally done these steps in pairs to represent the two possible states of a variable (such as p and p', etc.)

We have discussed various theoretical approaches to this, but up until now we have concentrated on the approach of using a set of three enzymes to do this writing. To write on DNA in this approach, it is cut first by a restriction enzyme that leaves overhanging ends of DNA at the cut site. Second, we use a DNA polymerase to fill in the overhang ends to make a blunt end. Then third we ligate those pieces back together using the third enzyme, DNA ligase. The DNA polymerase adds bases equal to the number of overhang nucleotides from the enzyme. This allows two things, first the site will no longer be cleaved by the original restriction enzyme and second it increases the size of the DNA by the number of added bases. Thus, this method writes on DNA and we can distinguish between written and un-written DNA by size and by its ability to be cleaved by a specific restriction enzyme. We implement the clauses for SAT problems on the DNA molecules by performing parallel digestions with specific restriction enzymes. In this problem, cleaved DNA does not satisfy the condition and is removed. If there is a molecule at the end that has remained uncut, it has satisfied all the conditions and is an answer to the problem. We read the answers for a problem by sequencing the DNA and/or by digesting it with the appropriate restriction enzymes. We have used this approach to address two classical problems in DNA computing namely a 3-variable SAT problem and a 3×3 Knight problem [2–5].

But for a variety of reasons, we wished to explore other writing approaches for aqueous computing. The three enzymes involved each require different enzyme conditions (buffers and salts) and so must be used sequentially and not simultaneously. This approach also requires the DNA be purified between steps to remove the unwanted buffer and salt components. These steps take significant time and result in some losses of DNA material at each step. To alleviate some of these challenges, we explore in this paper the use of DNA methylation as a modification that can be used for DNA computation. DNA methylases are naturally occurring enzymes in both prokaryotic and eukaryotic organisms that add a methyl group either to an adenosine or a cytosine (reviewed in [1] and [7]). In the former, they are believed to protect the organism from infection by a virus. These prokaryotes carry restriction enzymes to cut specific DNA sequences and a corresponding methylase that modifies the same DNA sequence in the host DNA. This modification blocks the restriction enzyme from cutting host DNA while incoming foreign DNA (say from a virus) does not have the methylation modification at that specific sequence and so is cleaved. This cleavage generally renders the viral DNA non-infectious, thus protecting the prokaryotic organism. Eukaryotic cells tend to use DNA methylation to affect gene regulation by DNA binding proteins. In most cases, methylated DNA is not expressed by eukaryotic cells. Because of the commercial availability of methylases and restriction enzymes, we thought it would be worthwhile to test whether this process could be used for DNA computing. Thus, here DNA methylation is the writing step while discerning whether a restriction enzyme can cut the DNA is part of the reading phase. The fact that one enzyme was going to be used at this writing step (the methylase) instead of 3, we felt would potentially speed up the computation. Thus, in this report, we tested 8 sets of methylases and restriction enzymes and did two 3-variable SAT problems with these enzymes. We used both plasmid and double-stranded PCR products for these reactions. We found this system works fairly well with some limitations in obtaining the correct answers to these problems.

2. Materials and Methods

The starting plasmid DNA, pBluescript SKII (Stratagene) was initially used and also the PCR product derived from amplification of the multiple cloning site of this plasmid with the M13-20 and reverse primers. Methylases were obtained either from New England Biolabs (BamHI methylase, dam methylase, EcoRI methylase, HaeIII methylase, HhaI methylase, and HpaII methylase) or Takara (HindIII methylase and ClaI methylase). Restriction enzymes and Taq polymerase were obtained from New England Biolabs. For methylation, generally 11–15 μg of DNA was incubated in 250 μl of the preferred enzyme buffer in the presence of 120 μM of S-adenosyl methionine and 24–48 units of the methylases. This was incubated for 12 hours at 37°C, heated at 65°C for 15 minutes and then the DNA purified using QIAquick PCR Purification Kit (Qiagen). Restriction enzyme digestions occurred with approximately 0.3 μg DNA in 10 μl of the enzyme's preferred buffer with 4–20 units of the restriction enzyme. The samples were incubated at the optimal temperature for the enzyme (37°C for all enzymes except SmaI which was incubated at 25°C and BssHII which was incubated at 50°C) for 1 hour. To confirm cleavage or not of the DNA, it was separated on an acrylamide (10% acrylamide Ready gel from BioRad) or agarose gel (0.8% in TBE buffer 0.09M Tris, 0.09M borate, 2mM EDTA) and visualized with ethidium bromide. Images of the gels were taken using a Kodak Imaging system.

3. Results

3.1. *Explanation of the system set-up and approach*

DNA methylases modify DNA so that certain restriction enzymes can no longer cleave the site. We wanted to test DNA methylation as a possible means for DNA computations in the aqueous style we have previously used [2, 4–6]. The overall scheme is in two stages for a standard 3-variable SAT problem (shown in Figure 1).

We first assign the six Boolean literals (p, p', q, q', r and r') specific sites on a given piece of double-stranded DNA (Figure 1A). These sites correspond to specific restriction enzymes sites much like we have done in other implementations [2, 4–6]. Initially, all of the sites can be cleaved by the restriction enzymes. We define a literal as being True if the site corresponding to this variable can be cleaved by the restriction enzyme. Thus, initially each literal is true which is not logically consistent (p and p' cannot both true). We created the collection of logically consistent DNA molecules by methylating the restriction enzyme sites to prevent them from being cleaved. We did this in pairs, each with half of the DNA (see Figure 1A where it is shown for variables p and p'). Thus, in half of the DNA, methylation is done on the p site while in the other half,

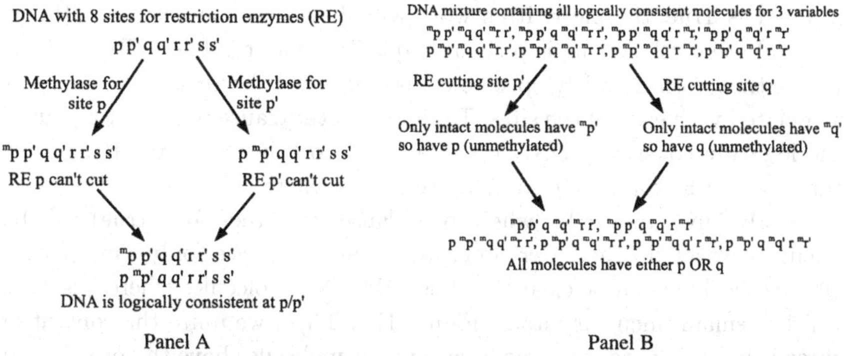

Figure 1. Representation of using methylation as a computation tool in a SAT problem. A. To create the logically consistent molecules at 4 variables we start with a plasmid DNA (or a PCR product) with 8 sites for distinct restriction enzymes (designated p, p', etc.). Half of these molecules are treated with the methylase for the p site and half treated with the methylases for the p' site. The action of each methylase blocks the cutting by the corresponding restriction enzyme. When the product of those two reactions are recombined, the resulting DNAs are methylated either at p or at p' and so are considered logically consistent at this variable. This is continued for the other 3 variables to make a mixture of DNAs that are logically consistent at all 4 variables. B. To implement the clauses, we cut with the appropriate restriction enzymes to destroy molecules that do not satisfy the clause. In the example given, to implement p OR q, we separate the result from part A with logically consistent molecules at 3 variables into two pots. In one, we cut with the restriction enzyme for the p' site which should leave intact only those molecules where the p' site is methylated. In the other pot, we cut with the restriction enzyme for the q' site which should leave intact only those molecules where the q' site is methylated. When these two are combined the intact molecules have either the p site unmethylated (p' methylated) or the q site unmethylated (q' methylated) and so for the intact molecules p OR q has been satisfied. We continue with this approach for the other clauses.

methylation is put on at the p' site. When these two parts are recombined together, no DNA molecules contain both p and p' sites unmethylated and thus cleavable. By doing this reaction, we have enforced logical consistency for the pair of literals p and p'. In the problem, we then continue with the methylases corresponding to the sites for the literals q and q' and then the literals r and r'. The final resulting DNA is logically consistent with regard to each individual Boolean variable and should contain 8 distinct types of DNA molecules.

In the second stage, we proceeded with various cleavage steps to invoke the chosen clauses of the SAT problem being tested. The problem we used here was to determine if there is a truth setting for the Boolean variables p, q, r for which each of the four clauses p OR q, p' OR q OR r', q' OR r', p' OR r is True. Simple written work will show that the answer to this is Yes and the truth setting is p is False, q is True and r is False. To weed out DNAs that do not satisfy specific clauses we used the restriction enzymes associated with each of the sites. To represent the clause p OR q, we poured the logically consistent DNA (above) into two tubes (Figure 1B). In one tube, we cut at the p' site (only those DNAs where p' is True will be cut), thus only DNA molecules where p' is False (p is therefore True) will be uncut or intact. In the other tube, we cut at the q' site (only those DNAs where q' is True will be cleaved), thus only DNA molecules where q is True will be remain uncut or intact (Figure 1B). Then, we unite the content of these tubes and note that, the intact, uncut molecules have the property of p OR q = T. We continued in a similar manner with each other clause. At the end of all the clauses, we determined if there is any intact, uncut DNA left and if so, what is the methylation status (and thus the designation) of each literal site. In this way, we hoped to be able to use methylation instead of the three-enzyme protocol we had used previously [2, 4, 5].

3.2. *Testing available methylases and restriction enzymes*

We found 8 commercially available methylases and the corresponding restriction enzymes that we used to confirm whether this system could be used as a binary modification on DNA and therefore a potential technique for DNA computations. These enzymes are listed in Table 1.

We first needed to confirm the specificity of each of the methylases for its target site and only that site on the DNA. For this experiment, we methylated the DNA first with a specific methylase, then we confirmed that the DNA was completely methylated using the target restriction enzyme. This was done by showing this restriction enzyme could not cut the DNA. Then, we cut the methylated DNA with all of the other restriction enzymes to confirm that these others could cleave this methylated DNA. All 64 combinations were tested (8 differently methylated DNA with the 8 different restriction enzymes), most both on plasmid and on PCR product. Checking of the dam methylase was only done on PCR product as the plasmid DNA was obtained from a dam positive host thus the DNA was already methylated at that site. In all cases, once completely methylated DNA was obtained, only the appropriate enzyme was blocked from cleaving the DNA. Some representative gels are shown in Figure 2.

We had at times difficulty in obtaining completely methylated DNA due to apparent low activity of some of the DNA methylases. We found that in

Table 1. Listing of methylase and restriction enzymes used for computation

Methylase	Seq. modified[1]	Restriction enzyme blocked & sequence recognized
BamHI methylase	G-G-A-T-^{m}C-C	BamHI GGATCC[2]
Dam methylase	G-^{m}A-T-C	DpnII GATC[2]
ClaI methylase	A-T-C-G-^{m}A-T	ClaI ATCGAT
EcoRI methylase	G-A-^{m}A-T-T-C	EcoRI GAATTC
HaeIII methylase	G-G-^{m}C-C	NotI GCGGCCGC
HhaI methylase	G-^{m}C-G-C	BssHII GCGCGC
HindIII methylase	^{m}A-A-G-C-T-T	HindIII AAGCTT
HpaII methylases	C-^{m}C-G-G	SmaI CCCGGG

[1]Sequence after methylase modification with the ^{m}A or ^{m}C indicating a methylated adenine or cytosine residue, respectively.
[2]Although the BamHI methylase and dam methylase overlap in their sequence specificity, the BamHI restriction enzyme is not inhibited by dam methylase modification while the DpnII restriction enzyme is not inhibited by the modification of the BamHI methylase. NdeI, Sau3A and MboI also cut at this site. MboI is not sensitive to methylation.

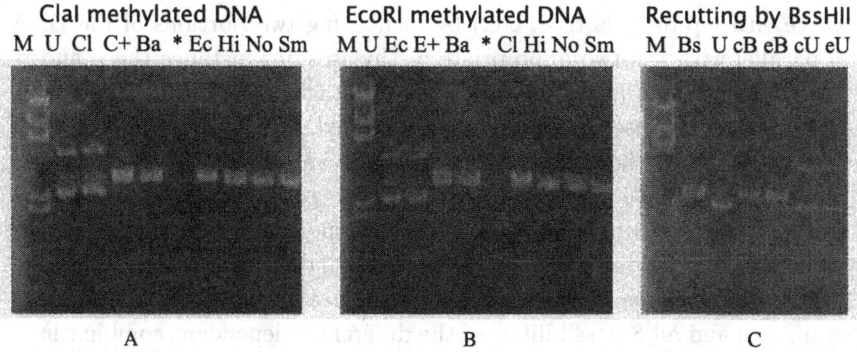

Figure 2. Confirmation that methylation does not affect the other restriction enzymes. Plasmid DNA was methylated with either ClaI methylase (panel A and C- labeled cB and cU) or with EcoRI methylase (panel B and C- labeled eB and eU), then left uncut (U,cU,eU) or cut with either ClaI (Cl), BamHI (Ba), EcoRI (Ec), HindIII (Hi), NotI (No), SmaI (Sm) or BssHII (cB,eB), then separated on a 0.8% agarose gel with a λ HindIII cut DNA marker (M). As a positive control, unmodified plasmid DNA was cut with Cla (C+), EcoRI (E+) or BssHII (Bs). Uncut DNA shows up as multiple bands on these gels representing open circular and supercoiled DNA while cut DNA shows up as one band at 3Kb.

some cases the DNA needed to be incubated with the methylase 2 or 3 times before the site was completely methylated (measured by complete lack of cleavage by the corresponding restriction enzyme). This was especially true of BamHI methylase and HindIII methylase. We tried longer incubations and higher concentrations of S-adenosyl methionine and methylases but it did not always make any difference in the outcome. Thus, we showed that each of the individual methylases modified only their corresponding site and did not disrupt the cleavage by restriction enzymes recognizing different sites, so we could begin with the computations.

3.3. *3-variable SAT computation with plasmid DNA and PCR product*

We tried two different schemes for doing the same satisfiability (SAT) problem with 3 variables and 4 clauses. In one scheme, we used circular plasmid DNA pBluescript that contains a multiple cloning site containing all the restriction enzyme/methylase recognition sites and in the other case, a PCR amplification product of this same region. The assignments of methylases for the individual Boolean variables for both the plasmid and PCR products are given in Table 2.

In both cases, we first created the self-consistent set of DNAs (ones where either p or p' is true, etc.) by incubating two portions of the DNA in parallel with the two appropriate methylases, for instance the p and p' methylase (scheme in Figure 1A). At each stage, we confirmed that each individual methylase had completely modified the DNA by showing the lack of cutting by the corresponding restriction enzyme before recombining the two halves. We then separated into two equal portions to continue with the other two methylases to represent the q and q' literals, followed with the methylation by the enzymes corresponding to the r and r' literals. In the final tube, we expect to have only logically consistent molecules (p or p', etc) and all 8 possibilities of the different independent combinations. Following the preparation of this pool of different DNAs, we then performed restriction enzyme cleavage steps to remove those DNAs that did not satisfy the clauses using the scheme described above and in Figure 1B. Following

Table 2. Boolean variable assignments used

	For plasmid		For PCR product
p	HindIII site	p	BssHII site
p'	SmaI site	p'	NotI site
q	NotI site	q	BamHI site
q'	BssHII site	q'	DpnII site
r	ClaI site	r	ClaI site
r'	EcoRI site	r'	EcoRI site

the reduction in complexity of the DNA molecules using the clauses, we determined whether there was an answer (intact, uncut DNA) and if so, what the truth settings for the answer was for the 3 Boolean variables.

For plasmid DNA, we had 3 different ways to look for intact DNA. We could look for uncut circular DNA as we had in the previous work (see [5]). Alternatively, we could look for DNA able to transform bacteria as the pBluescript plasmid contains an origin of replication for E. coli bacteria and an ampicillin resistance gene for selecting bacteria containing the plasmid and only circular, intact plasmid DNA will efficiently transform bacteria (see [5]). Finally, we could use PCR to amplify the full multiple cloning site using primers annealing to the two different sides of this region (M13 reverse primer and M13 -20 primer). Practically, we found too little of the plasmid DNA to detect on a gel directly but we did find enough circular DNA to transform bacteria (data not shown) and to produce a PCR product (Figure 3, lane 2).

Thus, there is an answer that satisfies all the clauses. To determine the status of each literal, we cleaved the plasmid DNA first with each of the six restriction enzymes, then performed PCR using the primers that were located on either side of the multiple cloning site. If a site is methylated, the enzyme should not cut (site is defined as False) and a PCR product should

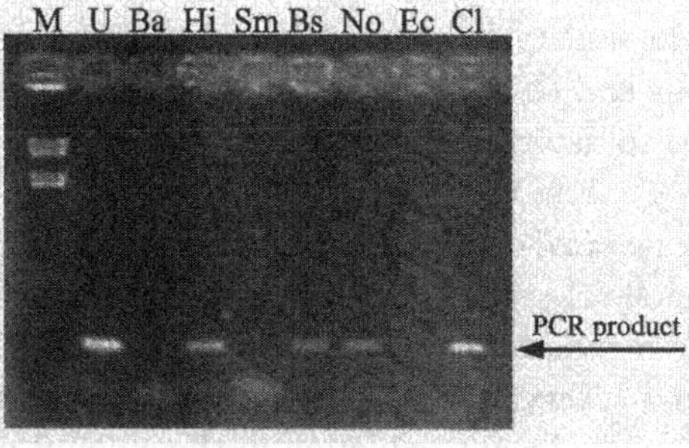

Figure 3. PCR product from treated plasmid DNA following SAT computation using methylation. Circular plasmid resulting from the SAT computation was either untreated (U) or cut with various restriction enzymes (BamHI=Ba, HindIII=Hi, SmaI=Sm, BssHII=Bs, NotI=No, EcoRI=Ec, ClaI=Cl) then amplified using PCR with M13 -20 and reverse primers and the product separated on a 1.5% agarose gel with the λ DNA cut with HindIII as molecular weight marker (M). If a PCR product is present, that indicates the plasmid is intact and the site was not cut (presumably due to the site being methylated) and therefore the site would be considered False. If there is no PCR product produced, that indicates the restriction site was cut and therefore the site would be designated True. Uncut DNA and BamHI cut DNA act as positive and negative controls, respectively.

be formed. If a site can be cleaved (non-methylated and is True), the sites for the two opposing PCR primers are separated from each other and no product is formed. We used this strategy to detect the status of the 6 sites involved (Figure 3). The final product of the clauses can act as a template for successful PCR to produce a 300 bp fragment. Cleavage by an enzyme found in the multiple cloning site but not used in the computation shows that cleavage blocks the ability to form a PCR product (Figure 3, lane 3). Then the figure shows the pairs of sites next to each other on the gel. It is clear that cleavage by HindIII and ClaI still allow the production of a PCR product suggesting those sites are methylated. On the other hand, cleavage by enzymes SmaI and EcoRI block the ability of the PCR to produce a product (Figure 3, lanes 5 and 8). These sites are thus defined as true. This makes both p and r to be False as SmaI and EcoRI represent p' and r', respectively. Unfortunately as can be seen in the figure, both NotI and BssHII cleavage result in some PCR product although in both cases the amount of product is decidedly less than that detected after HindIII or ClaI cleavage. The NotI and BssHII enzymes represent the literals q and q', respectively and thus we do not expect both to be false as represented by the data presented (Figure 3). The expected answer for the problem is p is False, q is True, and r is False. Wc found the correct settings for both p and r, but not for q where the results call both q and q' as False.

We also performed the satisfiability problem with PCR product from an amplification of the multiple cloning region containing all of the sites used (see above). At the end of the clauses, we separated the resulting DNA on an acryl-amide gel to determine if any full length PCR product still existed (Figure 4B, lane 6). The full-length PCR fragment was detected on the gel along with many smaller cleavage products that had not been removed from the mixture following the clauses. Next, to determine the status of each of the literals, we incubated the remaining DNA with different restriction enzymes followed by acrylamide gel electrophoresis (Figure 4). Cleavage by EcoRI results in significant reduction in the amount of full-length PCR product detected relative to cleavage by the alternative enzyme site, ClaI (Figure 4A). Likewise, cleavage by DpnII results in more full-length DNA than cleavage by the enzyme BamHI (Figure 4B, lanes 2 and 3). When the DNA is incubated with either NotI or BssHII, more of the full-length PCR product was detected after cleavage by the latter enzyme than the former although there is still significant full-length PCR product in the case of the sample treated with NotI. Also, still some full-length DNA is visible in the EcoRI and BamHI digested samples suggesting incomplete cleavage

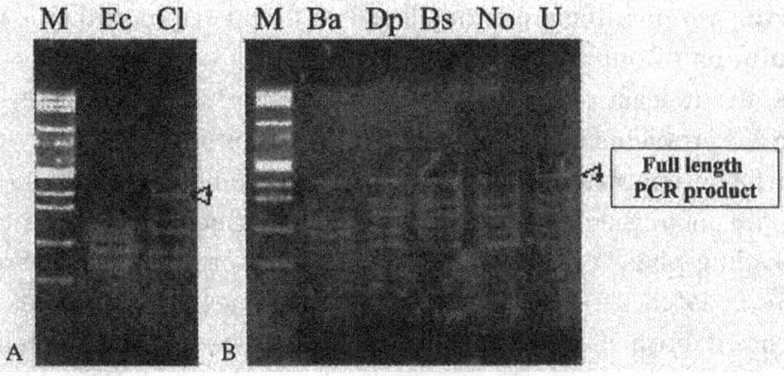

Figure 4. Analysis of DNA following SAT computation using methylation of PCR product. DNA fragments resulting from the SAT computation using PCR product as the initial template was either untreated (U) or cut with various restriction enzymes (EcoRI=Ec, ClaI=Cl, BamHI=Ba, DpnII=Dp, BssHII=Bs, NotI=No,) then separated on a 10% acrylamide gel with the θX174 DNA cut with HaeIII as molecular weight marker (M). The arrowheads mark the full length PCR product. If the full-length PCR product is present, that indicates the DNA is intact and the site was not cut (presumably due to the site being methylated) and therefore the site would be considered False. If there is less full-length PCR product than in the untreated sample, that indicates the restriction site was cut and therefore the site would be designated True. Uncut DNA acts as a positive control.

of the final product and potentially incompletely methylated DNA. Thus, through the literal assignments in Table 2, we come up with the answer p is False, q is True and r is False with the caveat that there is still a significant amount of uncut DNA in all samples.

4. Discussion

To analyze the possibility of using the bacterial DNA methylation system as a means for DNA computations, we first had to explore whether the methylase enzymes were functioning and specific, meaning they would alter only the recognition of their linked restriction enzyme and would not alter the recognition by any of the other restriction enzymes we used. This appeared to be the case. We found evidence that the DNA methylase enzymes were sequence specific and no evidence of cross-methylation to affect the other restriction enzyme sites used in this study. We then used these 8 methylases and 8 restriction enzymes in two DNA computations in the aqueous style of our previous work. With this approach, we were able to perform two 3-variable SAT problems using 4 clauses. In both cases, we found evidence for an answer, however, the answer provided contained two possible states for a single given variable. Simple analysis of the chosen clauses did not indicate two possible answers. Thus, we found some problems with the implementation of our approach.

There are at least three explanations for the lack of complete success of these SAT problems using the methylation approach, incomplete methylation during preparation of the logically consistent DNA set, incomplete digestion during implementation of the clauses or incomplete digestion during the reading phase of the computation. If the original DNA was incompletely methylated during the preparation, some DNAs in our mixture might contain both paired sites available for cleavage. As stated above, we often had problems getting completely methylated DNA during the preparation of the logically consistent DNAs and had to repeat the methylation step 2 and 3 times for two enzymes. We did not proceed with the computation unless we observed complete lack of cutting by the restriction enzyme associated with the methylation site. But, potentially if a small amount of incompletely methylated DNA were still present, we may not have seen it on our gels yet it may have carried through the clause steps. However, in this case, we would have expected no answer as more DNA would have been destroyed during the implementation of the clauses than should have been due to the lack of a methylated (protected) site. What we found were answers that had both True and False as states of the variable which is not as easy to explain through lack of complete initial methylation. One alternative explanation for the partial lack of success is less than complete digestion by the restriction enzymes during the implementation of the clauses. This would allow some DNA to persist that should have been removed during these steps. Related to this explanation is the one where incomplete cutting may have occurred during the reading phase of the computation after all the clause steps had been completed. Either of these explanations would likely be consistent with the appearance of a DNA answer that should not have been maintained in the population.

In both attempts at the SAT problem, the enzyme pair NotI and BssHII was associated with the variable having the two possible answers. The latter of these two enzymes requires cutting at 50°C while the other works at 37°C. During the implementation of the clauses, we digested the DNA with one or the other of these enzymes to destroy the appropriate DNA molecules. In reading the final answers, we digested with these two enzymes as well before either performing PCR (with the plasmid) or before running on an acrylamide gel (for the PCR product). In neither case, did we confirm complete digestion of the material during clause implementation potentially allowing a small bit of incompletely digested DNA to persist in the mixture.

In the end, we only assess the presence of complete, uncut DNA from the original mixture and if incompletely cut material has carried over from previous steps, we cannot easily distinguish this from the appropriate answer without imposing our own biases. In both cases where the answers obtained showed two possible states of the variable, we were using the methylases HhaI and HaeIII and the restriction enzymes BssHII and NotI to test these sites, respectively. We expected the final answer molecule in both cases to be methylated at the BssHII site and have the NotI site unmodified. However, the molecule with the reverse arrangement seemed also to be present in both cases. Both answers maintained an intact molecule with a BssHII site unmethylated when this should have been destroyed during the clauses or the reading. If at either point, the BssHII enzyme did not cut the material completely, it would appear to be methylated at this site. Again, due to the challenges with this restriction enzyme, we feel that either incomplete digestion by BssHII during the clause or reading phase may explain our lack of complete success. Other methylases or an isoschizomer for BssHII that cuts more reproducibly would help make this approach more practical.

We undertook this work to find an alternative to our 3-step, 3-enzyme reaction to modify the state of each variable to make logically consistent DNA molecules [2, 4, 5]. We hoped that the use of a single enzyme, namely the methylases, would significantly speed up our processing. This has happened. However, our new approach has a number of limitations. In some cases, the methylase enzymes were not as effective in modifying the DNA and we found we needed to repeat certain methylation steps 1 or 2 times before completely methylated DNA at a site was obtained. We were also limited by the amount of material left at the end of the computation in our procedure making detection sometimes difficult. With these reactions, it is not possible to amplify the DNA material before the end of the computation as the action of the Taq polymerase to copy the strands during PCR alleviates the methylation state at these sites. It would be necessary therefore, to start with enough material to obtain a clear answer at the end. With the necessity of re-methylating some of the samples 1 and 2 times, the loss of material became more acute. There may be a way to remethylate the required strands after PCR as some DNA modification enzymes have been isolated that will modify hemi-methylated DNA to make both strands methylated [8]. This would have to be done after each round of PCR to reinstate the methyl groups at the positions of the parent strands. This may make it possible to expand this approach to make it easier to detect the answer with a small amount of material. We did use PCR but only in the last reading step where we were ascertaining the existence of an intact plasmid molecule after cleavage. The methylation approach is also limited by the number of commercially available methylases and corresponding restriction enzymes. Our previous approach only required a restriction enzyme that left an overhang after cleavage that could be filled in through the action of a DNA polymerase [2, 4, 5]. There are many hundreds of these types of restriction enzymes that are available. In our new method, we used 8 methylase enzymes and their corresponding 6-base cutting restriction enzymes. Many more methylases are known but are not yet commercially available. Thus we have a difficult limitation to our technique. It is possible other DNA modifications or the use of DNA binding proteins to protect certain sites from methylation could be used in a further expansion of this technique. Thus, this advance may make it possible to expand this approach to a much larger number of Boolean operators in the future.

References

[1] Grace Goll, M. & T.H. Bestor. Eukaryotic cytosine methyltransferases, *Annual Review of Biochemistry* **74** (2005) 481–514.

[2] Head, T., M. Yamamura & S. Gal, Aqueous computation: writing on molecules, *Proceedings of the 1999 Congress on Evolutionary Computation* **2** (1999) 1006–1010.

[3] Head, T. & S. Gal, Aqueous computing: writing into fluid memory, *Bulletin of the European Association for Theoretical Computer Science* **75** (2001) 190–198.

[4] Head, T., X. Chen, M.J. Nichols, M. Yamamura & Gal, S., Aqueous solutions of algorithmic problems: emphasizing knights on a 3 × 3, in: *DNA Computing, Lecture Notes in Computer Science, LNCS* **2340**, N. Jonoska and N.C. Seeman., eds., Springer-Verlag, Berlin (2002a) 191–202.

[5] Head, T., X. Chen, M. Yamamura & S. Gal, Aqueous computing: a survey with an invitation to participate, *Journal of Computer Science and Technology* **17** (2002b) 672–681.

[6] Head, T. & S. Gal, Aqueous computing: writing on molecules dissolved in water, in: *Nanotechnology: Science and Computation*, J. Chen, N. Jonoska & G. Rozenberg, eds., Springer-Verlag, Berlin, (2006)321–334.

[7] Jeltsch, A., Beyond Watson and Crick: DNA methylation and molecular enzymology of DNA methyltransferases. *ChemBioChem* **3**(2002)274–93.

[8] Pradhan, S., A. Bacolla, R.D. Well, & R.J. Roberts. Recombinant human DNA (cytosine-5) methyltransferase. I. Expression, purification and comparison of de novo and maintenance methylation. *Journal of Biological Chemistry* **274**(1999)33002–33010.

DYNAMICS OF RANDOM BOOLEAN NETWORKS

JAMES F. LYNCH

Department of Mathematics and Computer Science
Box 5815, Clarkson University
Potsdam, NY 13699-5815, USA
jlynch@clarkson.edu

Boolean networks are models of genetic regulatory networks. S. Kauffman based many of his claims about spontaneous self-organization in complex systems on simulations of randomly constructed Boolean networks. Some of these claims are precise mathematical statements. We analyze these statements using combinatorial methods and show that there is partial agreement with some of Kauffman's conclusions, but in other cases there is disagreement. Our key finding is an algebraic parameter that determines the likelihood of ordered behavior in a random Boolean network. There is a threshold such that when the parameter is less than the threshold, ordered behavior is prevalent, and when it is greater than the threshold, chaotic behavior is highly likely. When the parameter equals the threshold, some forms of ordered behavior persist, but others do not.

PACS Codes: 87.10, 82.39.R, 02.10, 89.75.F

Keywords: Boolean networks, cellular metabolism, random graphs, stability.

1. Introduction

The realization that the genome is a dynamic network, where some genes regulate the activity of other genes, is not new. It can be traced as far back as the Nobel prize winning discovery of the lac operon by Jacob and Monod [11]. Many other regulatory genes have been discovered since then, and a picture of the genome as a complex web of interacting genes has gradually emerged. Although the importance of this regulatory web was appreciated, mainstream molecular biology has largely focused on achieving a thorough understanding of the mechanisms underlying the interactions between individual molecular species. Until relatively recently, little attention was given to the study of the global behavior of molecular systems. This has changed with the recognition of systems biology as an important emerging discipline.

Stuart Kauffman and René Thomas are among the first and most influential researchers to recognize the importance of the systems viewpoint in molecular biology. Much of their work pertains to Boolean networks. We will define Boolean networks more precisely later, but essentially they are highly simplified models of the genome, where each gene is modeled by a Boolean element called a gate. At any time, each gate is either active or inactive, i.e., its state is either 1 or 0. The structure of the network determines how the activity of combinations of gates affects the activity of other gates. This is only an approximation of the behavior of actual genomic networks, but it was a reasonable starting point for several reasons. First, precise knowledge of the speed of molecular reactions was unavailable. Second, even if such knowledge had been available, the complexity of simulating reactions with many different rates would have overwhelmed the capabilities of existing computers. Both of these problems still exist, although much more is known now, and computing power has greatly increased in the last few decades.

A third reason for considering Boolean networks is that Kauffman and Thomas were interested in general properties of genomic networks and not in the detailed modeling of specific systems. Again, it makes sense to study the simplest and most basic model in order to find properties that apply to all gene nets.

Kauffman was interested in certain emergent properties of Boolean networks that are sometimes referred to as "spontaneous order" or "order for free." To summarize his thesis briefly, the genome is constructed from unreliable parts that are subject to damage from the environment. Further, their function can change as a result of mutation during reproduction. Yet the genome behaves in a robust and reliable manner. Kauffman argued that this order and stability was not solely the result of natural selection. There had to be a statistical tendency toward order and self-organization. In other words, natural selection acts on self-organizing systems rather than creating them. Without this tendency toward order, almost all mutations would be fatal, thus preventing evolution through natural selection. Kauffman has written extensively on this subject; his thinking is consolidated in his book [14].

Much of the evidence for Kauffman's thesis comes from computer simulations. The typical experiment consisted of randomly constructing a Boolean network subject to some constraints. Each gate's state was randomly initialized to 0 or 1, and then the system ran synchronously. Since the network has a finite number of gates, each of which has two possible states, the network itself has a finite number of states, and it will eventually return to some state that it had been in previously. Since the network operates deterministically, it will keep returning to this state, repeating the same sequence of states indefinitely. This sequence of states is called the limit cycle.

According to Kauffman, the behavior of the network prior to entering its limit cycle is analogous to the behavior of an embryonic cell as it differentiates into its ultimate cell type, and the limit cycle is analogous to the differentiated cell's replication cycle.

Three measures of order were considered:

1. The number of weak gates, i.e., gates that can be perturbed without changing the limit cycle that the network enters.
2. The number of gates that eventually freeze, i.e., they eventually stop changing state.
3. The size of the limit cycle.

Ordered behavior is characteristic of genomic and other biological networks. They have a large proportion of weak gates and frozen gates and a small limit cycle. Weak gates are a form of robustness—the ability to recover from small perturbations or errors. The presence of frozen gates indicates a degree of predictability in the network's behavior. A small limit cycle is another form of predictability: the system will repeat itself frequently. The opposite kinds of behavior are characteristic of non-biological, chaotic systems. A system where many of the gates are not weak shows sensitivity to initial conditions. The limit cycles in chaotic systems are very large, with many elements changing state unpredictably. Thus they are similar to strange attractors in continuous chaotic dynamical systems.

Since not all Boolean networks show ordered behavior, a basic problem is to find properties of networks that determine whether they will be ordered or chaotic. The degree of interconnection appears to be very important in this regard. This was varied in Kauffman's experiments by specifying the number of inputs each gate had, i.e., the number of gates whose states directly affected the gate. This number k was fixed, and a random network was generated by choosing, for each gate, its k input gates and assigning a Boolean function of k arguments to the gate. From the computer simulations, networks where $k \geq 3$ were chaotic, while those where $k \leq 2$ were very stable in the three senses described above. This is a kind of phase transition, where a change in microscopic conditions causes a change in macroscopic behavior.

Living cells seem to exhibit all three kinds of ordered behavior, and the typical gene is affected by only a few other genes. Thus the simulations provide evidence of spontaneous order in cells. In fact, Kauffman's computer simulations seemed to show a mathematical relationship between the number of gates and the size of the limit cycle in stable Boolean networks, which was paralleled in living cells by the number of genes and the duration of the replication cycle. The size of the limit cycle (or the duration of the replication cycle) appeared to be on the order of the square root of the number of gates (respectively genes).

One possible explanation for stability in Boolean networks where the gates had at most two inputs was that there are only 16 Boolean functions of two arguments, and two of those functions are constant, i.e., the function that always outputs 1 regardless of the values of its inputs, and the function that always outputs 0. Thus a significant fraction (1/8) of the gates in a random Boolean network with $k = 2$ would be constant, and this could be the source of global stability of the network. Conversely, the chaotic behavior observed in nets with $k \geq 3$ could be a consequence of the much smaller proportion of constant gates. However, Kauffman also ran simulations of randomly constructed networks with $k = 2$ but without constant gates, where the remaining 14 two argument functions were equally likely, and the results were similar to those where all 16 functions were used.

Kauffman proposed another category of functions as the source of order. He called these the canalyzing functions. A canalyzing function is a Boolean function for which there exists some argument and some Boolean value such that the output of the function is determined if the argument has that value. For example, the 2-argument OR function $x_1 \vee x_2$ is canalyzing because if either argument has the value 1, then the value of $x_1 \vee x_2$ is 1. Fourteen out of the sixteen 2-argument Boolean functions, including the constant functions, are canalyzing, but this proportion drops rapidly among Boolean functions with more than two arguments. Thus the hypothesis that nets with many canalyzing gates tend to be ordered, while those with few of them tend not to be, is consistent with the experimental results.

The sudden change in behavior between random Boolean networks with $k \leq 3$ and those with $k \leq 2$ is also evidence of the "edge of chaos" phenomenon that has been observed in many complex systems. The idea is that very stable and highly ordered systems are too simple and rigid to adapt to unpredictable influences from the environment; on the other hand, chaotic systems lack the robustness to maintain favorable states. Complex adaptive systems such as living cells are able to survive because they exist in a region between excessive stability and chaos. This theme has been elaborated on by Kauffman and many other researchers in complex systems, for example P. Bak [1] C. Langton [15], N. Packard [22], and S Wolfram [25].

All of these definitions and claims have precise mathematical formulations, so a natural question is whether the experimental results are supported by proofs. Indeed, this has been proven for the two extremes in the range of k. Networks with $k = 1$ are highly ordered in the three senses above (S. Jaffe [12]). Networks with $k = n$, where n is the number of gates, are equivalent to random functions on a set of size 2^n. It can be shown that in this case, most gates will not be weak or frozen, and a classical result of B. Harris [9] implies that average limit cycle size is $(\sqrt{\pi/8})2^{n/2}$.

Interestingly, at about the same time that Kauffman started investigating random Boolean networks, the mathematical techniques for dealing with random networks were being developed by P. Erdős and A. Rényi [6, 7] and E. Gilbert [8], but it was about 30 years before any of these techniques were applied to the analysis of random Boolean networks. The first proofs of any of Kauffman's claims for networks with $1 < k < n$ appear in an article co-authored by the mathematical biologist J. Cohen and the random graph theorist T. Łuczak [4].

Random graph theory is now a flourishing branch of combinatorics. The most extensively studied version of random graph is the independent edge model. In this version, there is a probability p (which may depend on the number of vertices in the graph) such that for each pair of vertices independently, there is an undirected edge between them with probability p. Graph theorists have discovered many deep and interesting results about this kind of random graph, but it does not seem to be a good model of the random networks studied in biology, communications, and engineering. A major distinction is that the degree distribution of this kind of graph is Poisson, but the degree distributions of many real-world networks obey a power law. A better model for these situations may be random graphs with a specified degree distribution, which are considered in recent articles by M. Molloy and B. Reed [19, 20]. Some other shortcomings of the standard version of random graph pointed out by M. Newman, S. Strogatz, and D. Watts [21] are that it is undirected and has only one type of vertex. They develop some techniques for dealing with random directed graphs with vertices of several types. However, even this model lacks the structure needed to model the dynamic behavior of networks.

Kauffman's Boolean networks are a further extension of the models in [21] that do include this additional structure. The gates of a Boolean network are vertices assigned a type corresponding to a Boolean function, and the directed edges indicate the inputs to each gate. But instead of simply regarding each vertex as a static entity, we are interested in how the functions of the gates change the state of the network over time. Our random Boolean networks include Kauffman's networks as a special case. They are specified by a sequence of probabilities $p_1, p_2, \ldots$, whose sum is 1, where for each gate independently, p_i is the probability that it is assigned the ith Boolean function. (We are assuming some canonical ordering of the finite Boolean functions.) Once each gate has been assigned its function, its indegree is determined by the number of arguments of the function, and its input gates are chosen at random using the uniform distribution. Lastly, a random initial state is chosen.

Our main result is an algebraic parameter, derived from the distribution $p_1, p_2, \ldots$, whose value determines the global behavior of the network. When the parameter is less than or equal to a certain threshold, ordered behavior of the first two kinds mentioned above is highly likely: almost all gates freeze quickly, and almost all gates can be perturbed without affecting the long-term behavior of the network. Conversely, if the parameter is larger than the threshold, the networks do not behave in such an ordered fashion. Our condition for stability actually implies forms of ordered behavior stronger than Kauffman's. That is, the gates freeze in time on the order of $\log n$, where n is the number of gates, and the effect of a perturbation dies out within order $\log n$ steps. Consequently, the failure of our condition implies forms of disordered behavior that are weaker than the negations of Kauffman's.

We then apply our main results to the two classes of 2-input Boolean networks mentioned above. Here, our analysis verifies Kauffman's claims for networks that use all 16 of the 2-argument Boolean functions, but it casts doubt on similar claims for networks that use only the 14 nonconstant functions and the importance of canalyzing functions as a source of order.

Our techniques can also be applied to the limit cycle size. Here the picture is not as complete. However, it does provide the only case of an analytic result which definitely contradicts the experimental conclusions. Using the same parameter as above, when it is strictly less than the threshold, limit cycles are small. But when the parameter equals the threshold, limit cycles suddenly get very large. The analysis is quite involved and the details will be presented in another article. We sketch these results at the end of this article. We do not know the behavior of limit cycle sizes when the parameter is greater than the threshold.

Slightly weaker versions of the results in this article were presented in [17] and [18]. The main contribution of this article is a more general and uniform treatment of these results.

2. Definitions

A Boolean network B is a 3-tuple $\langle V, E, \mathbf{f} \rangle$ where V is a set $\{1, \ldots, n\}$ for some natural number n, E is a set of labeled directed edges on V, and $\mathbf{f} = (f_1, \ldots, f_n)$ is a sequence of Boolean functions such that for each $v \in V$, the number of arguments of f_v is $\text{indeg}(v)$, the indegree of v in E, i.e., the number of edges entering v. These edges are labeled $1, \ldots, \text{indeg}(v)$. The interpretation is that V is a collection of Boolean gates, E describes their interconnections, and $\mathbf{f}$ describes their operation.

The gates update their states synchronously at discrete time steps $0, 1, \ldots$. At any time t, each gate v is in some state $x_v \in \{0, 1\}$. Letting $\mathbf{x} = (x_1, \ldots, x_n)$, we say that B is in state $\mathbf{x}$ at time t. Let $\text{indeg}(v) = m$ and $u_1, u_2, \ldots, u_m$ be the gates such that for $i = 1, \ldots, m$, $(u_i, v) \in E$ with label i. These are referred to as the *in-gates* of v. Then the state of v at time $t+1$ is $y_v = f_v(x_{u_1}, \ldots, x_{u_m})$. Letting $\mathbf{y} = (y_1, \ldots, y_n)$, we put $B(\mathbf{x}) = \mathbf{y}$.

The following notation will be used to describe the dynamical properties of Boolean networks.

Definition 1. Let $\mathbf{x} \in \{0,1\}^n$, the set of sequences of n 0's and 1's indexed by $1, \ldots, n$.

1. For $t = 0, 1, \ldots$, we put $B^t(\mathbf{x})$ for the state of B at time t, given that its state at time 0 is $\mathbf{x}$. That is,

$$B^0(\mathbf{x}) = \mathbf{x}, \quad \text{and}$$
$$B^{t+1}(\mathbf{x}) = B(B^t(\mathbf{x})) \text{ for all } t.$$

   We also put $B_v^t(\mathbf{x})$ for y_v where $\mathbf{y} = B^t(\mathbf{x})$.
2. Gate v freezes to $y \in \{0,1\}$ in t steps on input $\mathbf{x}$ if $B_v^{t'}(\mathbf{x}) = y$ for all $t' \geq t$.

3. For $\mathbf{x} \in \{0,1\}^n$ and $v \in \{1,\ldots,n\}$, we put $\mathbf{x}^v$ for the state which is identical to $\mathbf{x}$ except that $x_v^v = 1 - x_v$.
4. Let $u, v \in \{1,\ldots,n\}$ and $\mathbf{x} \in \{0,1\}^n$. We say that v affects u at time t on input $\mathbf{x}$ if $B_u^t(\mathbf{x}) \neq B_u^t(\mathbf{x}^v)$. We put

$$A_+^t(v,\mathbf{x}) = \{u \in V : v \text{ affects } u \text{ at time } t \text{ on input } \mathbf{x}\} \quad \text{and}$$
$$A_-^t(v,\mathbf{x}) = \{u \in V : u \text{ affects } v \text{ at time } t \text{ on input } \mathbf{x}\}.$$

5. Gate v is t-ineffective on input $\mathbf{x}$ if $A_+^t(v,\mathbf{x}) = \emptyset$, i.e., $B^t(\mathbf{x}) = B^t(\mathbf{x}^v)$.

Note that if v is t-ineffective for some t, then it is weak. In analyzing the robustness of Boolean networks, we will estimate the number of t-ineffective gates, for suitable t, since this appears more tractable than estimating the number of weak gates. Similarly, instead of analyzing the frozen gates, we will consider a stronger property of gates.

We will be examining randomly constructed Boolean networks. The random model we use is sufficiently general to capture the particular classes of random Boolean networks in the literature. Let $\phi_1, \phi_2, \ldots,$ be some ordering of all the finite Boolean functions, and let $p_1, p_2, \ldots,$ be a sequence of probabilities such that $\sum_{i=1}^{\infty} p_i = 1$. The selection of a random Boolean network with n gates is a three stage process. First, each gate is independently assigned a Boolean function using the distribution $p_1, p_2, \ldots$. That is, for each $v = 1,\ldots,n$ and $j = 1,2,\ldots,$ the probability that gate v is assigned ϕ_j is p_j. The probabilities may depend on n, the number of gates in the network; that is, each probability is actually a function $p_i(n)$. For example, $p_i(n) = 0$ for any ϕ_i with more than n arguments. For simplicity of notation, we suppress the functional notation. Next, the in-gates for each gate are selected. If the gate has been assigned an m-argument function, then its in-gates are chosen from the $n(n-1)\cdots(n-m+1)$ equally likely possibilities. Finally, random initial states are independently chosen for each gate.

We make several restrictions on the distribution $p_1, p_2, \ldots,$ still consistent with the random networks in the literature. We assume that the average and variance of the number of arguments of a randomly selected Boolean function, or equivalently, the average and variance of the indegree of a gate, are finite. That is, letting each ϕ_i have m_i arguments, $\sum_{i=1}^{\infty} p_i m_i^2 < \infty$. Our methods require that the state of a random gate be independent of the time, i.e., for any gate v, the probability that $B_v^t(\mathbf{x}) = 1$ is a constant. This is ensured in the following way.

Definition 2. For any natural number m and $\mathbf{x} \in \{0,1\}^m$, let $v(\mathbf{x}) = |\{i : 1 \leq i \leq m \text{ and } x_i = 1\}|$. Then for any $a \in [0,1]$,

$$\rho(a) = \sum_{i=1}^{\infty} p_i \sum_{j=1}^{m_i} \sum_{\substack{\mathbf{x} \in \{0,1\}^{m_i} \\ \phi_i(\mathbf{x})=1}} a^{v(\mathbf{x})}(1-a)^{m_i - v(\mathbf{x})} \tag{1}$$

is the probability that a random gate's state is 1 at time 1, given that each of its in-gates has probability a of being in state 1 at time 0.

Putting $\rho^{(t)}(a) = \rho(\rho(\cdots(a)\cdots))$ (ρ iterated t times), we require that there exists $a \in [0,1]$ such that $\lim_{t\to\infty} \rho^{(t)}(b) = a$, where b is the probability that a random gate's state is 1 at time 0. This condition is easily seen to be satisfied by the random Boolean networks in the literature. They have $b = \frac{1}{2}$, and $a = \frac{1}{2}$ since every Boolean function and its negation have equal probability of being assigned to a gate.

This class of random Boolean networks includes as special cases Kauffman's networks, networks with the classical random graph topology with edge probability cn^{-a}, $a \geq 1$ [2], networks with power law degree distribution $\propto d^{-c}$, $c > 1$, smallworld networks ([23]) and many of their variations ([21], [24]).

If θ is a property of boolean nets then $\text{pr}(\theta)$ denotes the probability that a random boolean network with n gates satisfies θ. If ϕ is also a property then $\text{pr}(\theta \mid \phi)$ is the probability of θ over random boolean networks with n gates, conditioned on ϕ.

3. Local Structure of Networks

For small intervals of time, the dynamical properties of the network are determined by its "local" structure. Thus, the gates affected by a given gate over the time interval $0, 1, \ldots, t$ lie in the portion of the network consisting of all gates reachable from the gate by a path in E of length at most t. Similarly, the behavior of a gate over the same interval is determined by the portion consisting of all gates that can reach the gate by such a path. Of course, for large enough t, these portions will be the entire network. The next definitions capture these notions of locality.

Definition 3. 1. For any subset $I \subseteq V$,

$$S_+^0(I) = I, \text{and}$$
$$S_+^{t+1}(I) = \{\, u : (v,u) \in E \text{ for some } v \in S_+^t(I) \,\} \text{ for } t \geq 0.$$

That is, $S^t_+(I)$ is the set of gates at the ends of paths of length t that start in I. Similarly, $S^t_-(I)$ is the set of gates at the beginnings of paths of length t that end in I.

2. Then

$$N^t_+(I) = \bigcup_{s=0}^{t} S^s_+(I), \quad \text{and}$$

$$N^t_-(I) = \bigcup_{s=0}^{t} S^s_-(I)$$

are the out- and in-neighborhoods respectively of I of radius t.

We put $S^t_+(v)$ for $S^t_+(\{v\})$ and similarly for the other notations. Thus the state of gate v at time t is determined by the states of the gates in $S^t_-(v)$ at time 0 and the functions assigned to the gates in $N^{t-1}_-(v)$.

We put log for $\log_2$. For the remainder of this article, α and β will be positive constants satisfying $2\alpha \log \delta + 2\beta < 1$ and $\alpha \log \delta < \beta$, where $\delta = \mathbf{E}(m_i)$.

Lemma 1. *Let $I \subseteq \{1, \ldots, n\}$, $|I| \leq n^\beta$, and $t \leq \alpha \log n$. The following events have probability $1 - o(1)$:*

1. *For every $v \in I$, $N^t_+(v)$ induces a tree in $\langle V, E \rangle$.*
2. *For every distinct $u, v \in I$, $N^t_+(u) \cap N^t_+(v) = \emptyset$.*
3. *For every $v \in I$, $N^t_-(v)$ induces a tree in $\langle V, E \rangle$.*
4. *For every distinct $u, v \in I$, $N^t_-(u) \cap N^t_-(v) = \emptyset$.*

Proof. The lemma follows by showing that each of these events fails with probability $o(1)$. The calculations are similar for all events, and we show the work only for event 1.

If 1. fails, then there exist distinct gates $v_1, \ldots, v_s$ such that

$s \leq \alpha \log n$,
for $i = 1, \ldots, s-1$, v_i is an in-gate of v_{i+1}, and
$v_1 \in I$,

and distinct gates $w_1, \ldots, w_r$ such that

$r \leq \alpha \log n$,
for $i = 1, \ldots, r-1$, w_i is an in-gate of w_{i+1},
$w_1 \in \{v_1, \ldots, v_s\}$,

$\{w_2, \ldots, w_r\} \cap \{v_1, \ldots, v_s\} = \emptyset$, and
there is a labelled edge (w_r, v_s) distinct from the labelled edge (v_{s-1}, v_s).

Now s and r can be chosen in $O((\log n)^2)$ ways. The gates $v_1, \ldots, v_s$ and $w_1, \ldots, w_r$ can be chosen in $O(n^{s+r-2+\beta} \log n)$ ways. For each $i = 1, \ldots, s-2$, the probability that v_i is an in-gate of v_{i+1} is

$$\sum_{i=1}^{\infty} p_i \frac{m_i}{n} = \frac{\delta}{n}.$$

Similarly, the probability that each w_i is an in-gate of w_{i+1} for $i = 1, \ldots, r-1$ is δ/n. The probability that both v_{s-1} and w_r are in-gates of v_s is

$$\sum_{i=1}^{\infty} p_i \frac{m_i(m_i - 1)}{n(n-1)} = O(n^{-2}).$$

Altogether, the probability that 1. fails is

$$\begin{aligned} O\big((\log n)^3 \times n^{s+r-2+\beta} \times (\delta/n)^{s+r-3} \times n^{-2}\big) &= O\big((\log n)^3 \delta^{2\alpha \log n} n^{\beta-1}\big) \\ &= O\big((\log n)^3 n^{2\alpha \log \delta + \beta - 1}\big) \\ &= o(1). \end{aligned}$$ □

4. Branching Processes

As we showed, for t not large compared to n, the typical $N_+^t(v)$ induces a tree in a Boolean network with n gates. A perturbation of the state of such v may cause perturbations to the states of $S_+^1(v)$ in the next step, then $S_+^2(v)$, and so on, in a "wave" that propagates through $N_+^t(v)$. It is possible that this wave dies out and the effects of the perturbation are transient, i.e., gate v is t-ineffective. We will show that this behavior can be approximated by a branching process. Then, by applying basic results about branching processes, we will derive our results about ineffective gates. Similarly, $N_-^t(v)$ induces a tree for almost all v. In this case, the gates that affect v at times $0, \ldots, t$ are the successive generations in a branching process that propagates backwards through $N_-^t(v)$.

We will summarize the results about branching processes that we need. For more information on branching processes, see T. Harris [10].

A branching process can be identified with a rooted labelled tree. The tree may have infinite branches. Each node will be labelled with the unique path from the root to that node. That is, the root is labelled with the null sequence. If the root has k children, they are labelled with the sequences

(1), (2), ..., (k). If the second child of the root has l children, then they are labelled with the sequences $(2,1)$, $(2,2)$, ..., $(2,l)$, and so on. Generation t consists of all nodes labelled with a sequence of length t. The number of children of any node is independent of the number of children of any other node, but the probability of having a certain number of children is the same for all nodes. Thus the probability space of a branching process is determined by a sequence $(q_k : k = 0, 1, \dots)$ where q_k is the probability that a node has k children. The probability measure on this space will be denoted by bpr. In describing events in this space, P will denote a branching process. If χ is a property of branching processes, $P \models \chi$ means χ holds for P, and $\text{bpr}(P \models \chi)$ is the probability that χ holds.

For $t \geq 0$, $P \restriction t$ will be the finite labelled tree which is P restricted to its first t generations. Z_t will be the random variable which is the size of generation t, i.e., the number of nodes of depth t.

The generating function of the branching process is the series

$$F(z) = \sum_{k=0}^{\infty} q_k z^k.$$

That is, $F(z)$ is the probability generating function of Z_1 since $q_k = \text{bpr}(Z_1 = k)$. A basic result is that the t-th iterate of $F(z)$ is the probability generating function of Z_t. The iterates of $F(z)$ are defined by

$$F_0(z) = z \quad \text{and}$$
$$F_{t+1}(z) = F(F_t(z)) \quad \text{for } t \geq 0. \tag{2}$$

Then

Theorem 1. *The probability generating function of Z_t is $F_t(z)$, i.e.,*

$$F_t(z) = \sum_{k=0}^{\infty} \text{bpr}(Z_t = k) z^k.$$

This enables us to express the moments of Z_t in terms of the moments of Z_1, which in turn have simple representations in terms of the derivatives of $F(z)$. Let μ and σ^2 be the first and second moments of Z_1, that is, $\mu = \mathbf{E}(Z_1)$ and $\sigma^2 = \text{var}(Z_1)$.

Theorem 2. *We have*

$$\mu = F'(1) \quad \textit{and}$$
$$\sigma^2 = F''(1) + F'(1) - (F'(1))^2.$$

More generally, for all $t \geq 0$, the first and second moments of Z_t are

$$\mathbf{E}(Z_t) = \mu^t \quad \textit{and}$$

$$\operatorname{var}(Z_t) = \begin{cases} \dfrac{\sigma^2 \mu^t(\mu^t - 1)}{\mu^2 - \mu} & \textit{if } \mu \neq 1, \\ t\sigma^2 & \textit{if } \mu = 1. \end{cases}$$

Our analysis of frozen and ineffective gates uses two branching processes, both of which are defined in terms of a parameter that characterizes the tendency of gates to be influenced (regulated) by their in-gates.

Definition 4. Let $f(x_1, \ldots, x_m)$ be a Boolean function of m arguments, and $\mathbf{x} = (x_1, \ldots, x_m) \in \{0, 1\}^m$ be an assignment of 0's and 1's to its arguments. For $i \in \{1, \ldots, m\}$, we say that argument i directly affects f on input $\mathbf{x}$ if $f(\mathbf{x}) \neq f(\mathbf{x}^i)$. We put $\gamma(f, \mathbf{x})$ for the number of i's that directly affect f on input $\mathbf{x}$. We extend this notion to gates in a Boolean network in the obvious way. Given a Boolean network B where gate v has in-gates $u_1, \ldots, u_m$ and state $\mathbf{x} \in \{0, 1\}^n$, for $i = 1, \ldots, m$, u_i directly affects v on input $\mathbf{x}$ if $B_v(\mathbf{x}) \neq B_v(\mathbf{x}^{u_i})$.

Let

$$\lambda = \sum_{i=1}^{\infty} p_i \sum_{\mathbf{x} \in \{0,1\}^{m_i}} \gamma(\phi_i, \mathbf{x}) a^{v(\mathbf{x})} (1 - a)^{m_i - v(\mathbf{x})}. \tag{3}$$

Thus λ may be regarded as the average number of arguments that directly affect a random Boolean function with a random input.

5. Ineffective Gates

As we showed, for sufficiently small I and t, the "typical" $N_+^t(I)$ and $N_-^t(I)$ induce a forest on $\langle V, E \rangle$, i.e., there are no directed or undirected cycles among their gates. If this is the case for $N_+^t(v)$, then we can give a simple recursive definition of $A_+^t(v, \mathbf{x})$.

Lemma 2. *Assume $N_+^t(v)$ induces a tree on E. Then for any $s \leq t$, any $\mathbf{x} \in \{0, 1\}^n$, and any gate $u \in S_+^s(v)$, v affects u at time s on input $\mathbf{x}$ if and only if*

1. *$s = 0$ and $u = v$, or*
2. *$s > 0$ and, letting w be the unique gate such that $w \in S_+^{s-1}(v) \cap S_-^1(u)$, v affects w at time $s - 1$ on input $\mathbf{x}$, and w directly affects u on input $B^{s-1}(\mathbf{x})$.*

We will use the branching process defined as follows.

$$q_k = \frac{\lambda^k}{k!}e^{-\lambda}$$

for $k = 0, 1, \ldots$. Therefore $F(z) = e^{\lambda z - \lambda}$. From Theorem 2,

$$\begin{aligned}
\mu &= \lambda, \\
\sigma^2 &= \lambda, \\
\mathbf{E}(Z_t) &= \lambda^t, \quad \text{and} \\
\mathrm{var}(Z_t) &= \begin{cases} \dfrac{\lambda^t(\lambda^t - 1)}{\lambda - 1} & \text{if } \mu \neq 1, \\ t\lambda & \text{if } \mu = 1. \end{cases}
\end{aligned}$$

Definition 5. Let T be a labelled tree of height t, $B = \langle V, E, \mathbf{f} \rangle$ be a Boolean network, and $\mathbf{x} \in \{0, 1\}^n$ be its state. For $v \in \{1, \ldots, n\}$, we put $T \Rightarrow v$ if

$N^t_-(A^t(v, \mathbf{x}))$ induces a tree in $\langle V, E \rangle$, and
there is an isomorphism from T onto $\langle A^t_+(v, \mathbf{x}), E \rangle$.

Lemma 3. *If* $|T| \leq n^\beta$ *and the height of* T *is* $t \leq \alpha \log n$, *then for all* $\mathbf{x} \in \{0, 1\}^n$, $\mathrm{pr}(T \Longrightarrow v) = \mathrm{bpr}(P \restriction t \cong T)(1 + o(1))$.

Proof. By Lemma 1, if there is an isomorphism τ from T onto $\langle A^t_+(v, \mathbf{x}), E \rangle$, then almost surely $N^t_-(A^t_+(v, \mathbf{x}))$ induces a tree in $\langle V, E \rangle$. Thus we need only analyze the probability that τ exists. Let $u_1, \ldots, u_h$ be the non-leaf nodes of T, in lexicographic order. The construction of τ is recursive and proceeds in stages $1, \ldots, h$. At each stage s, $\tau(u_s)$ has been defined at some previous stage, and it is extended to the children of u_s. (At stage 1, $\tau(u_1) = v$ has already been defined.) Also, the Boolean functions assigned to these children are selected.

Thus, assume that at stage s, $\tau(u_1), \ldots, \tau(u_{K_s})$ have already been defined, where $s \leq K_s$. Let u_s have k_s children. Then there are $\binom{n-K_s}{k_s}$ ways of selecting the children of $\tau(u_s)$ in $\langle V, E \rangle$. Having chosen these children, we next assign Boolean functions to them. Independently, for each child w of $\tau(u_s)$, let ϕ_i be assigned to it. This event has probability p_i, and for $j = 1, \ldots, m_i$, the probability that $\tau(u_s)$ is the jth in-gate of w is $1/n$. Summing over all i, we get the probability that $\tau(u_s)$ directly affects w:

$$\sum_{i=1}^{\infty} p_i \sum_{\mathbf{x}\in\{0,1\}^{m_i}} \sum_{j=1}^{m_i} \mathrm{pr}(\tau(u_s)\text{is the } j\text{th in-gate of } w, \text{ the initial state is } \mathbf{x},$$

$$\text{and } u_s \text{ directly affects } w \text{ on input } \mathbf{x} \mid f_w = \phi_i)$$

$$= \sum_{i=1}^{\infty} \frac{p_i}{n} \sum_{\mathbf{x}\in\{0,1\}^{m_i}} \gamma(\phi_i, \mathbf{x}) a^{v(\mathbf{x})} (1-a)^{m_i - v(\mathbf{x})}$$
$$= \frac{\lambda}{n}.$$

Therefore the probability that these k_s gates are directly affected by $\tau(u_s)$ is $(\lambda/n)^{k_s}$.

Since the events of assigning Boolean functions to all the gates are independent, the probability that the selected gates belong to $A_+^t(v, \mathbf{x})$ is

$$\prod_{s=1}^{h} \binom{n-K_s}{k_s} \left(\frac{\lambda}{n}\right)^{k_s} = \left(\prod_{s=1}^{h} \frac{\lambda^{k_s}}{k_s!}\right) \left(1 - \frac{O(n^\beta)}{n}\right)^{O(n^\beta)}$$
$$= \left(\prod_{s=1}^{h} \frac{\lambda^{k_s}}{k_s!}\right) (1 - O(n^{2\beta-1})).$$

The probability that no other gates are in $A_+^t(v, \mathbf{x})$ is

$$\left(1 - \frac{\lambda h}{n}\right)^{n-|T|} = e^{-\lambda h}(1 + O(n^{2\beta-1})).$$

Therefore

$$\mathrm{pr}(T \Rightarrow v) = \left(\prod_{s=1}^{h} \frac{\lambda^{k_s}}{k_s!} e^{-\lambda}\right) (1 + o(1))$$
$$= \mathrm{bpr}(P \restriction t \cong T)(1 + o(1)). \qquad \square$$

We say that a property χ of branching processes depends only on the first t generations if, for any two branching processes P_1 and P_2 such that $P_1 \restriction t \cong P_2 \restriction t$, either $P_1 \models \chi$ and $P_2 \models \chi$, or $P_1 \not\models \chi$ and $P_2 \not\models \chi$. Thus χ can be identified with a set of labelled trees of depth at most t. We will also use the notation $\langle A_+^t(v, \mathbf{x}), E\rangle \models \chi$ to mean $\langle A_+^t(v, \mathbf{x}), E\rangle$ induces a tree in $\langle V, E\rangle$ whose corresponding branching process satisfies χ.

Theorem 3. *Let χ be a property of branching processes that depends only on the first $\alpha \log n$ generations. Then for all $\mathbf{x} \in \{0,1\}^n$*

$$\mathrm{pr}(\langle A_+^t(v,\mathbf{x}), E\rangle \models \chi) = \mathrm{bpr}(P \models \chi) + o(1).$$

Proof. By the previous lemma, it suffices to show that $\mathrm{bpr}(|P \restriction \alpha \log n| \geq n^\beta) = o(1)$.

If $|P \restriction \alpha \log n| \geq n^\beta$, then $Z_t \geq n^\beta/(\alpha \log n)$ for some $t = 1, \ldots, \alpha \log n$. Since $\mathbf{E}(Z_t) = \lambda^t \leq \delta^t \leq n^{\alpha \log \delta} \ll n^\beta/(\alpha \log n)$,

$$\mathrm{pr}(Z_t \geq n^\beta/(\alpha \log n)) \leq \frac{\mathrm{var}(Z_t)}{\left(n^\beta/(\alpha \log n) - \mathbf{E}(Z_t)\right)^2} \text{ by Chebyshev's inequality}$$

$$= \begin{cases} \dfrac{\lambda^{2t-1} + \lambda^{2t-2} + \cdots + \lambda^t}{\left(n^\beta/(\alpha \log n) - \lambda^t\right)^2} & \text{if } \lambda \neq 1 \\ \dfrac{t\lambda}{\left(n^\beta/(\alpha \log n) - \lambda^t\right)^2} & \text{if } \lambda = 1 \end{cases}$$

$$= o(1/\log n) \text{ in either case.} \qquad \square$$

A gate v such that $N_-^{\alpha \log n}(A_+^{\alpha \log n}(v,\mathbf{x}))$ is acyclic is $\alpha \log n$-ineffective if and only if its corresponding branching process is extinct within $\alpha \log n$ generations. Clearly this depends only on the first $\alpha \log n$ generations, so Theorem 3 applies. By basic results from branching process theory, the probability of extinction in t generations is $\mathrm{bpr}(Z_t = 0) = F_t(0)$, and $\lim_{t\to\infty} F_t(0) = r$, where r is the smallest nonnegative root of $z = F(z)$. Further, when $\mu \leq 1$, $r = 1$, and when $\mu > 1$, $r < 1$. Therefore

Theorem 4. *There is a constant r such that for all $\mathbf{x} \in \{0,1\}^n$*

$$\lim_{n\to\infty} \mathrm{pr}(v \textit{ is } \alpha \log n\textit{-ineffective}) = r.$$

When $\lambda \leq 1$, $r = 1$, and when $\lambda > 1$, $r < 1$.

Corollary 1. *The expected number of $\alpha \log n$-ineffective gates in a random Boolean network is asymptotic to rn.*

A stronger result is

Corollary 2. *The number of $\alpha \log n$-ineffective gates in almost all Boolean networks is asymptotic to rn.*

That is, there is a function $\varepsilon(n)$ such that $\varepsilon(n) \to 0$ and, letting the random variable X_n be the number of $\alpha \log n$-ineffective gates in a random Boolean network with n gates,

$$\lim_{n\to\infty} \mathrm{pr}(|X_n - rn| \leq n\varepsilon(n)) = 1.$$

Proof. By the previous corollary,

$$\mathbf{E}(X_n) = rn + n\varepsilon(n),$$

where $\varepsilon(n)$ is a function such that $\lim_{n\to\infty} \varepsilon(n) = 0$. When $\lambda \leq 1$, $r = 1$, so, letting the random variable $Y_n = n - X_n$, by Markov's inequality

$$\mathrm{pr}\left(Y_n \geq n\sqrt{|\varepsilon(n)|}\right) = O\left(\sqrt{|\varepsilon(n)|}\right).$$

Therefore the corollary holds for $\lambda \leq 1$.

When $\lambda > 1$, $r < 1$, and we need to estimate $\mathrm{var}(X_n)$. Using methods similar to those in the proofs of Lemma 1 and Theorems 3 and 4 it can be shown that, for any two distinct gates u and v, almost surely $N_-^{\alpha \log n}(A_+^{\alpha \log n}(u, \mathbf{x}))$ and $N_-^{\alpha \log n}(A_+^{\alpha \log n}(v, \mathbf{x}))$ are acyclic, their intersection is empty, and

$$\lim_{n\to\infty} \mathrm{pr}(u \text{ and } v \text{ are } \alpha \log n\text{-weak }) = r^2.$$

Therefore

$$\mathrm{var}(X_n) = r(1-r)n + n^2\varepsilon'(n)$$

for some function $\varepsilon'(n) \to 0$. By Chebyshev's inequality

$$\mathrm{pr}(|X_n - rn - n\varepsilon(n)| > n\sqrt[4]{|\varepsilon'(n)|}) \leq \frac{r(1-r)n + n^2\varepsilon'(n)}{n^2\sqrt{|\varepsilon'(n)|}} \to 0,$$

and the corollary also holds for $\lambda > 1$. □

When $\lambda > 1$, it is also true that most of the $\alpha \log n$-effective gates affect many other gates when perturbed.

Corollary 3. *Let $\lambda > 1$. For almost all random Boolean networks, if gate v is $\alpha \log n$-effective, then there is a positive W such that for $t \leq \alpha \log n$, the number of gates affected by v at time t is asymptotic to $W\lambda^t$.*

Proof. For $t \geq 0$, let $W_t = Z_t/\mu^t$ $(= Z_t/\lambda^t$ in our case). Again by basic results from branching process theory, there is a random variable W such that

$$\mathrm{bpr}(\lim_{t\to\infty} W_t = W) = 1 \quad \text{and}$$
$$\lim_{t\to\infty} \mathrm{bpr}(Z_t \neq 0 \text{ and } W = 0) = 0. \tag{4}$$

From this the corollary follows. □

6. Frozen Gates

Estimating the number of gates that freeze seems to be quite difficult. However, there is a condition on gates whose in-neighborhoods are trees that implies freezing and which is amenable to combinatorial analysis.

For the remainder of this section, t will represent a natural number in the range $0, \ldots, \alpha \log n$, and y will be a variable taking on the values 0 and 1. Given a Boolean function $\phi(x_1, \ldots, x_m)$ and $\mathbf{x} = (x_1, \ldots, x_m) \in \{0, 1, *\}^m$, we say that $\mathbf{x}$ forces ϕ to y if, for all $\mathbf{x}' \in \{0, 1\}^m$ such that $x_i = x_i'$ whenever $x_i \neq *$, $\phi(\mathbf{x}') = y$. The $*$'s are "don't care" values, meaning their value does not affect the value of ϕ whenever the remaining arguments agree with $\mathbf{x}$. For example, ϕ is forced by every $\mathbf{x} \in \{0, 1\}^m$; if ϕ is a constant function, then it is forced by every $\mathbf{x} \in \{0, 1, *\}^m$; if $\phi(x_1, x_2) = x_1 \vee x_2$, then it is forced to 0 by $(0, 0)$ and to 1 by $(0, 1)$, $(1, 0)$, $(1, 1)$, $(1, *)$, and $(*, 1)$. We can now give a recursive definition of forcing for the gates of a Boolean network.

Definition 6. A gate v is forced to y in 0 steps if f_v is the constant function y.

For $t \geq 0$, v is forced to y in $t + 1$ steps if, letting $u_1, \ldots, u_m$ be its in-gates, there is $\mathbf{x} \in \{0, 1, *\}^m$ such that $\mathbf{x}$ forces f_v to y and for each $i = 1, \ldots, m$ such that $x_i \neq *$, f_{u_i} is forced to x_i in t steps. We say that v is forced (in some number of steps) if it is forced to 0 or 1.

It is clear that forcing is a stronger condition than freezing.

Lemma 4. *If a gate in a Boolean network is forced to y in t steps, then it freezes to y in t steps.*

When the in-neighborhood of a gate is a tree, forcing is related to the branching process where the children of a gate are its in-gates that directly affect it. The next lemma states that forcing is equivalent to extinction in this branching process. We put $A_-^t(v) = \bigcup_{\mathbf{x}} A_-^t(v, \mathbf{x})$.

Lemma 5. *If $N_-^t(v)$ induces a tree, then v is forced in t steps if and only if $A_-^t(v) = \emptyset$.*

Proof. The "only if" direction of the proof is immediate from the definitions.

To prove the other direction, we use induction on t. When $t = 0$, it is again immediate from the definitions.

Now assume the result holds for t, $N_-^{t+1}(v)$ induces a tree, and v is not forced in $t+1$ steps. Let $u_1, \ldots, u_m$ be as in Definition 6. Let $\mathbf{x} \in \{0, 1, *\}^m$ be defined by

$$x_i = \begin{cases} \text{forced value of } u_i & \text{if } u_i \text{ is forced in} t \text{steps} \\ * & \text{otherwise.} \end{cases}$$

Since v is not forced in $t+1$ steps, there is some j such that $x_j = *$ and $\mathbf{x}' \in \{0,1\}^m$ such that $x_i = x'_i$ whenever $x_i \neq *$ and $f_v(\mathbf{x}') \neq f_v(\mathbf{x}'^j)$. Since $N_-^{t+1}(v)$ induces a tree, by the induction hypothesis, $A_-^t(u_j) \neq \emptyset$. Then there are states $\mathbf{y}$ and $\mathbf{y}'$ such that

$$B_{u_i}^t(\mathbf{y}) = B_{u_i}^t(\mathbf{y}') = x'_i \quad \text{for } i \neq j, \text{ and}$$
$$B_{u_j}^t(\mathbf{y}) \neq B_{u_j}^t(\mathbf{y}').$$

But then $B_v^{t+1}(\mathbf{y}) \neq B_v^{t+1}(\mathbf{y}')$, and $A_-^{t+1}(v) \neq \emptyset$. □

We now use another branching process to model the propagation of gates that affect a given gate. The theorems about forced gates are similar to those about ineffective gates, but the proofs are simpler. Therefore we will only sketch them. This time we define the branching process by:

$$q_k = \sum_{i:m_i \geq k} p_i \sum_{\substack{x \in \{0,1\}^{m_i} \\ \gamma(\phi_i, \mathbf{x}) = k}} a^{v(\mathbf{x})}(1-a)^{m_i - v(a)}$$

for $k = 0, 1, \ldots$. Then $\mu = \lambda$ and by our conditions that $\sum_{i=1}^{\infty} p_i m_i^2 < \infty$, $\sigma^2 < \infty$. Again,

$$\mathbf{E}(Z_t) = \lambda^t, \quad \text{and}$$
$$\operatorname{var}(Z_t) = \begin{cases} \dfrac{\sigma^2 \lambda^t (\lambda^t - 1)}{\lambda^2 - \lambda} & \text{if } \mu \neq 1, \\ t\sigma^2 & \text{if } \mu = 1. \end{cases}$$

Definition 7. Let T be a labelled tree of height t, $B = \langle V, E, \mathbf{f} \rangle$ be a Boolean network, and $\mathbf{x} \in \{0, 1\}^n$ be its state. For $v \in \{1, \ldots, n\}$, we put $T \Rightarrow v$ if

$A_-^t(v, \mathbf{x})$ induces a tree in $\langle V, E \rangle$, and
there is an isomorphism from T onto $\langle A_-^t(v, \mathbf{x}), E \rangle$.

Lemma 6. *If $|T| \leq n^\beta$ and the height of T is $t \leq \alpha \log n$, then for all $\mathbf{x} \in \{0,1\}^n$, $\operatorname{pr}(T \Rightarrow v) = \operatorname{bpr}(P \restriction t \cong T)(1 + o(1))$.*

Theorem 5. *Let χ be a property of branching processes that depends only on the first $\alpha \log n$ generations. Then for all $\mathbf{x} \in \{0,1\}^n$*

$$\mathrm{pr}(\langle A_-^t(v,\mathbf{x}), E\rangle \models \chi) = \mathrm{bpr}(P \models \chi) + o(1).$$

Theorem 6. *There is a constant r such that for all $\mathbf{x} \in \{0,1\}^n$*

$$\lim_{n\to\infty} \mathrm{pr}(v \textit{ is forced in } \alpha \log n \textit{ steps}) = r.$$

When $\lambda \leq 1$, $r = 1$, and when $\lambda > 1$, $r < 1$.

Corollary 4. *The expected number of gates forced in $\alpha \log n$ steps in a random Boolean network is asymptotic to rn.*

Corollary 5. *The number of gates forced in $\alpha \log n$ steps in almost all Boolean networks is asymptotic to rn.*

That is, there is a function $\varepsilon(n)$ such that $\varepsilon(n) \to 0$ and, letting the random variable X_n be the number of gates forced in $\alpha \log n$ steps in a random Boolean network with n gates,

$$\lim_{n\to\infty} \mathrm{pr}(|X_n - rn| \leq n\varepsilon(n)) = 1.$$

When $\lambda > 1$, it is also true that most of the gates that are not forced in $\alpha \log n$ steps are affected by many other gates.

Corollary 6. *Let $\lambda > 1$. For almost all random Boolean networks, if gate v is not forced in $\alpha \log n$ steps, then there is a positive W such that for $t \leq \alpha \log n$, the number of gates that affect v at time t is asymptotic to $W\lambda^t$.*

7. Networks of 2-Input Gates

Since Kauffman's nets are special cases of our networks, a natural question is whether our analysis agrees with the conclusions of his experiments. We will apply our results to the two classes of 2-input random Boolean networks that Kauffman studied. In one case, we have agreement: the networks are stable with high probability. In the other case, our theorems do not contradict the experimental results, but they provide evidence that these networks are less stable than the first kind. Further, the conjecture that random Boolean networks with a high proportion of canalyzing gates are stable is not supported by our analysis. To explain these results, we use the notion of canalyzing to classify the 2-argument Boolean functions. A Boolean function $f(x_1, \ldots, x_m)$ is canalyzing if it is forced by some $\mathbf{x} \in \{0, 1, *\}^m$

where $x_i \neq *$ for exactly one $i \in \{1, \ldots, m\}$. We have three categories of 2-argument Boolean functions:

I. The two constant functions:

$$f(x_1, x_2) = 0 \quad \text{and} \quad f(x_1, x_2) = 1$$

II. The twelve nonconstant canalyzing functions, consisting of

A. The four functions that depend on one argument:

$$\begin{array}{lcl} f(x_1, x_2) = x_1 & \text{and} & f(x_1, x_2) = \neg x_1 \\ f(x_1, x_2) = x_2 & \text{and} & f(x_1, x_2) = \neg x_2 \end{array}$$

B. The eight canalyzing functions that depend on both arguments:

$$\begin{array}{rcl} x_1 \vee x_2 & \text{and} & \neg x_1 \wedge \neg x_2 \\ \neg x_1 \vee x_2 & \text{and} & x_1 \wedge \neg x_2 \\ x_1 \vee \neg x_2 & \text{and} & \neg x_1 \wedge x_2 \\ \neg x_1 \vee \neg x_2 & \text{and} & x_1 \wedge x_2 \end{array}$$

III. The two noncanalyzing functions exclusive or and equivalence:

$$x_1 \oplus x_2 \quad \text{and} \quad x_1 \equiv x_2$$

Since each function is paired with its negation, Equation 1 is satisfied with $a = 1/2$. Let p_{I}, p_{II}, and p_{III} be the respective sums of the probabilities of the functions of type I, II, and III, i.e., p_{I} is the probability that a gate is assigned a function of type I, and so on. We can now express the λ parameter of Section 4 (see Equation (3)) in terms of p_{I}, p_{II}, and p_{III}. Since $a = 1/2$, $a^{v(\mathbf{x})}(1-a)^{2-v(\mathbf{x})} = 1/4$ for every $\mathbf{x} \in \{0,1\}^2$. Clearly, if ϕ_i is of type I,

$$\gamma(\phi_i, \mathbf{x}) = 0 \text{ for every } \mathbf{x} \in \{0,1\}^2.$$

If ϕ_i is of type II.A., say $\phi_i(x_1, x_2) = x_1$, then

$$\gamma(\phi_i, \mathbf{x}) = 1 \text{ for every } \mathbf{x} \in \{0,1\}^2.$$

If ϕ_i is of type II.B., say $\phi_i(x_1, x_2) = x_1 \vee x_2$, then

$$\sum_{\mathbf{x} \in \{0,1\}^2} \gamma(\phi_i, \mathbf{x}) = 4.$$

Altogether, the type I and II functions contribute p_{II} to λ. Lastly, it is easily seen that if ϕ_i is a type III function, then

$$\gamma(\phi_i, \mathbf{x}) = 2 \quad \text{for every } \mathbf{x} \in \{0,1\}^2,$$

and therefore the type III functions contribute $2p_{\mathrm{III}}$ to λ, giving

$$\lambda = p_{\mathrm{II}} + 2p_{\mathrm{III}}.$$

By Corollaries 2 and 5, for almost all Boolean networks, almost all gates are $\alpha \log n$-ineffective if and only if $\lambda \leq 1$, and for almost all Boolean networks, almost all gates are forced in $\alpha \log n$ steps if and only if $\lambda \leq 1$. Since $\lambda = p_{\mathrm{II}} + 2p_{\mathrm{III}}$ and $p_{\mathrm{I}} + p_{\mathrm{II}} + p_{\mathrm{III}} = 1$, $\lambda \leq 1$ is equivalent to $p_{\mathrm{I}} \geq p_{\mathrm{III}}$. Therefore both types of ordered behavior hold if and only if $p_{\mathrm{I}} \geq p_{\mathrm{III}}$.[a]

Kauffman performed extensive simulations on two classes of random networks constructed from 2-argument Boolean functions. In the first class, all 16 of these functions were equally likely to be assigned to a gate. In the second, no constant functions were used, and the remaining 14 functions were equally likely. In the first case, $p_{\mathrm{I}} = p_{\mathrm{III}} = 1/8$. Therefore in this case, almost all gates are $\alpha \log n$-ineffective and forced in $\alpha \log n$ steps. But in the second case, $p_{\mathrm{I}} = 0$ and $p_{\mathrm{III}} = 1/7$, so $p_{\mathrm{I}} < p_{\mathrm{III}}$. Thus in this case, a nontrivial fraction of the gates are $\alpha \log n$-effective and not forced in $\alpha \log n$ steps. This does not directly contradict the experimental conclusions because it is possible that, even though the effect of a perturbation persists for $\alpha \log n$ steps, it could die out after that. And even if it persisted indefinitely, it may not change the limit cycle that the network enters. Similarly, a gate may not be forced, but it could still freeze.

8. Future Work and Open Problems

In [16], it was shown that, for random Boolean networks with 2-input gates, limit cycles are bounded in size with high probability when $p_{\mathrm{I}} > p_{\mathrm{III}}$. But in [17], it was shown that limit cycles are large at the threshold $p_{\mathrm{I}} = p_{\mathrm{III}}$. Specifically, the average size of the limit cycle is greater than n^a for any a. This disagrees with Kauffman's claim that the average limit cycle size is on the order of $\sqrt{n}$.

These results generalize to the classes of random Boolean networks considered in this paper: when $\lambda < 1$, limit cycles are bounded in size with high probability, but when $\lambda = 1$, the average size of the limit cycle is superpolynomial in the number of gates. It is reasonable to conjecture that the limit cycle size is monotonic in λ, and that limit cycles would be very large when $\lambda > 1$. This would also agree with the experimental evidence. However, we do not know of a proof.

[a] Article [17] contains proofs that $p_{\mathrm{I}} \geq p_{\mathrm{III}}$ implies these kinds of ordered behavior; it was conjectured in [17] that they fail when $p_{\mathrm{I}} < p_{\mathrm{III}}$.

Other conjectures about the long-term behavior of nets with $\lambda > 1$ are unresolved. Based on the results of this article, one would expect that the other forms of instability would be prevalent: sensitivity to initial conditions and many unfrozen gates.

From what we have shown, $\lambda = 1$ appears to be the point at which complex behavior begins. This idea could be investigated further by considering Boolean networks with inputs and outputs. These are an easy generalization of the networks studied here. Such networks can compute functions of their inputs, and we conjecture that when $\lambda < 1$, these functions are very simple with high probability, but when $\lambda = 1$, the functions can be artibrarily complex in the senses of computational complexity.

References

[1] P. Bak, *How Nature Works: The Science of Self-organized Criticality*, Oxford University Press (1997).

[2] B. Bollobás, *Random Graphs*, Academic Press (1985).

[3] J. M. Bower and H. Bolouri, *Computational Modeling of Genetic and Biochemical Networks*, The MIT Press, Cambridge, MA (2001).

[4] J. E. Cohen and T. Łuczak, Stability of vertices in random Boolean cellular automata, *Random Structures and Algorithms* **2** (1991) 327–334.

[5] B. Derrida and Y. Pomeau, Random networks of automata: a simple annealed approximation, *Europhys. Lett.* **1** (1986) 45–49.

[6] P. Erdős and A. Rényi, On random graphs, *Pub. Math.* **6** (1959) 290–297.

[7] P. Erdős and A. Rényi, On the evolution of random graphs, *Magyar Tud. Akad. Mat. Kutato Int. Kozl.* **5** (1960) 17–61.

[8] E. Gilbert, Random plane networks, *J. SIAM* **9** (1961) 533–543.

[9] B. Harris, Probability distributions related to random mappings, *Ann. Math. Stat.* **31** (1960) 1045–1062.

[10] T. E. Harris, *The Theory of Branching Processes*, Dover Publications, Inc., New York (1989).

[11] F. Jacob and J. Monod, Genetic repression, allosteric inhibition ad cellular differentiation, in *Cytodifferentiation and Macromolecular Synthesis*, M. Locke, ed., Academic Press, New York (1963) 30–64.

[12] S. Jaffe, Kauffman networks: Cycle structure of random clocked Boolean networks, Ph. D. diss., New York University (1988).

[13] S. A. Kauffman, Metabolic stability and epigenesis in randomly connected nets, *J. Theoret. Biol.* **22** (1969) 437–467.

[14] S. A. Kauffman, *The Origins of Order: Self-Organization and Selection in Evolution*, Oxford University Press, New York, (1993).

[15] C. G. Langton, Computation at the edge of chaos: phase transitions and emergent computation, *Physica D* **42** (1990) 12–37.

[16] A criterion for stability in random Boolean cellular automata, *Ulam Quart.* **2** (1993) 32–44.

[17] J. F. Lynch, On the threshold of chaos in random Boolean cellular automata, *Random Structures and Algorithms* **6** (1995) 239–260.
[18] J. F. Lynch, Critical points for random Boolean networks, *Physica D* **172** (2002) 49–64.
[19] M. Molloy and B. Reed, A critical point for random graphs with a given degree sequence, *Random Structures and Algorithms* **6** (1995) 161–179.
[20] M. Molloy and B. Reed, The size of the giant component of a random graph with a given degree sequence, *Combinatorics, Probabililty and Computing* **7** (1998) 295–305.
[21] M. E. J. Newman, S. H. Strogatz, and D. J. Watts, Random graphs with arbitrary degree distributions and their applications, *Phys. Rev. E* **64** (2001) 026118.
[22] N. H. Packard, Adaptation toward the edge of chaos, in *Dynamic Patterns in Complex Systems*, J. A. S. Kelso, A. J. Mandell, and M. F. Shlesinger, eds., World Scientific, Singapore (1988) 293–301.
[23] D. J. Watts and S. H. Strogatz, Collective dynamics of "small world" networks, *Nature* **393** (1998) 440–442.
[24] D. J. Watts, *Small Worlds: The Dynamics of Networks Between Order and Randomness*, Princeton University Press, Princeton, NJ (1999).
[25] S. Wolfram, *A New Kind of Science*, Wolfram Media Inc., Champaign, IL (2002).

UNKNOTS AND DNA

LOUIS H. KAUFFMAN
Department of Mathematics, Statistics and Computer Science
University of Illinois at Chicago
851 South Morgan St., Chicago IL 60607-7045, U.S.A.
kauffman@math.uic.edu

S. LAMBROPOULOU
Department of Mathematics
National Technical University of Athens
Zografou Campus, GR-157 80 Athens, Greece.
sofia@math.ntua.gr

0. Introduction

This paper gives infinitely many examples of unknot diagrams that are *hard*, in the sense that the diagrams need to be made more complicated by Reidemeister moves before they can be simplified. In order to construct these diagrams, we prove theorems characterizing when the numerator of the sum of two rational tangles is an unknot. The paper then uses these results in studying processive DNA recombination and finding minimal size unknot diagrams. This paper is a short version of a paper in which we include complete proofs of all statements. Many proofs are omitted in the present paper.

See Figure 2 for a diagram that we shall refer throughout this paper as the "Culprit." This culprit is not the only culprit, but it is the exemplar that we shall use, and it is the example that started this investigation. The first author likes to use the Culprit as an example in introductory talks about knot theory. One draws the Culprit on the board and asks whether it is knotted or not. This gives rise to a discussion of easy and hard unknots, and how the existence of hard unknots makes us need a theory of knots in order to prove knottedness when it occurs. After using this example, we began to ask how to produce other examples that were hard and to wonder if our familiar culprit might be the smallest such example (size being the number of crossings, in this case 10).

We show that there are infinitely many examples of hard unknot diagrams, obtained by using the theory of rational tangles and their closures. In order to use the theory of rational tangles, one must become familiar with the notion of *tangle* and the notion of the *fraction of a tangle.* In Section 1 we introduce the tangle analysis and assume that the reader knows about tangle fractions. We discuss the theory of tangle fractions in Section 2.

In Section 1 we see that the Culprit can be divided into two rational tangles whose fractions add up to a fraction whose numerator has absolute value equal to 1. It turns out that *whenever the sum of the fractions of two rational tangles has numerator equal to plus or minus one, then the closure of the sum of the two tangles will be an unknot.* This result is Theorem 5 in Section 3. In Theorem 7 of Section 4 we take a further step and characterize fractions $\frac{P}{Q}$ and $\frac{R}{S}$ such that $\frac{P}{Q} - \frac{R}{S} = \frac{\pm 1}{QS}$ in terms of their associated continued fractions. It turns out that this last equation is satisfied if and only if one of the two continued fractions is a *convergent* of the other. This means that one continued fraction is a one-term truncate of the other. For example

$$\frac{3}{2} = 1 + \frac{1}{2}$$

is a convergent of

$$\frac{10}{7} = 1 + \frac{1}{2 + \frac{1}{3}}.$$

Section 2 sets up the matrix representations for continued fractions that underpin the proof of the Theorems. This completely solves the question of when two fractions give rise to an unknot via the (numerator) closure of the sum of their associated tangles.

In Section 5 we use these results to construct many examples of hard unknots. The first example, $K = N([1,4] - [1,3])$, of this section is given in Figure 15 and its mirror image H in Figure 19. This culprit K is a hard unknot diagram with only 9 crossings. We then show how our original culprit (of 10 crossings) arises from a "tucking construction" applied to an unknot that is an easy diagram without the tuck (Figure 17). This section then discusses other applications of the tucking construct. In Section 6 we prove that the 9 crossing examples of Figure 19 and some relatives obtained by flyping and taking mirror images are the smallest hard unknot diagrams that can be made by taking the closure of the sum of two alternating rational tangles. In Section 7 we show the historically first hard unknot, due to Goeritz in 1934. The Goeritz diagram has 11 crossings.

In Section 8, we show how our unknots are related to the study of processive recombination of DNA. In the *tangle model* for DNA recombination, pioneered by DeWitt Sumners and Claus Ernst, the initial substrate of the DNA is represented as the closure of the sum of two rational tangles. It is usual to assume that the initial DNA substrate is unknotted. We have characterized such unknot configurations in this paper, and so are in a position to apply our results to the model. We show that processive recombination stabilizes, in the sense that the form of the resulting knotted or linked DNA is obtained by just adding twists *in a single site* on the closure of a certain tangle. This result helps to understand the form of the recombination process.

1. Culprits

Combinatorial knot theory got its start in the hands of Kurt Reidemeister [31] who discovered a set of moves on planar diagrams that capture the topology of knots and links embedded in three dimensional space. Reidemeister proved that the set of diagrammatic moves shown in Figure 1 generate isotopy of knots and links. That is, he showed that if we have two knots or links in three dimensional space, then they are ambient isotopic if and only if corresponding diagrams for them can be obtained, one from the other, by a sequence of moves of the types shown in Figure 1.

Here is an example of a knot diagram (originally due to Ken Millett [27]), in Figure 2. We like to call this diagram the "Culprit." The Culprit is a knot diagram that represents the unknot, but as a diagram, and using only the Reidemeister moves, it must be made more complicated before it can be simplified to an unknotted circle. We measure the complexity of a knot or link diagram by the number of crossings in the diagram. Culprit has 10

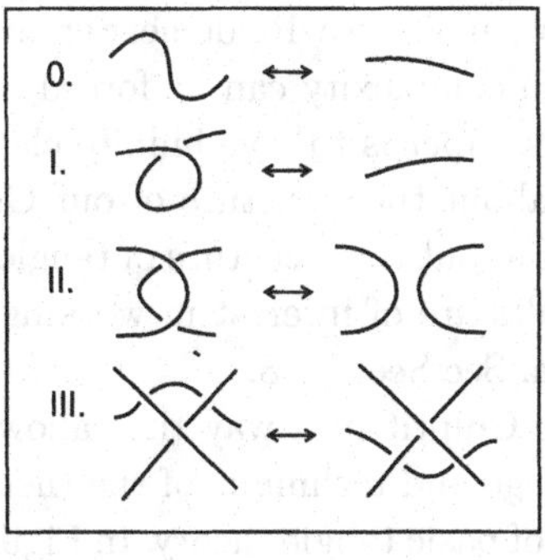

Figure 1. The Reidemeister moves.

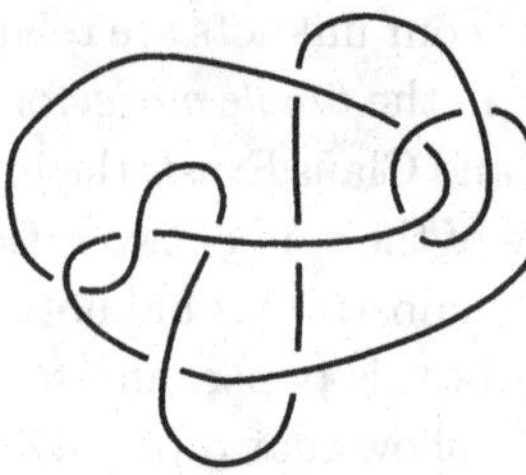

Figure 2. The culprit.

crossings, and in order to be undone, we definitely have to increase the number of crossings before decreasing them to zero. The reader can verify this for himself by checking each region in the diagram of the Culprit. A simplifying Reidemeister II move can occur only on a two-sided region, but no two-sided region in the diagram admits such a move. Similarly on the Culprit diagram there are no simplifying Redeimeister I moves and there are no Reidemeister III moves (note that a III move does not change the complexity of the diagram). We view the diagram of the Culprit and other such examples as resting on the surface of the two-dimensional sphere. Thus the outer region of the diagram counts as much as any other region in this search for simplifying moves.

We shall call a diagram of the unknot *hard* if it has the following three properties:

1. There are no simplifying Type I moves on the diagram.
2. There are no simplifying Type II moves on the diagram.
3. There are no Type III moves on the diagram.

Hard unknot diagrams have to be made more complex before they will simplify to the unknot, if we use Reidemeister moves. It is an unsolved problem just how much complexity can be forced by a hard unknot.

One purpose of this paper is to give infinite classes of hard unknots by employing an insight about the structure of our Culprit, and generalizing this insight into results about the structure of tangles whose numerators are unknotted. These results are of interest in working with the tangle model of DNA recombination. See Section 8.

In order to see the Culprit in a way that allows us to generalize him, we shall use the language and technique of the theory of tangles. The next sections describe a bit of basic tangle theory. In Figure 3 analyze the Culprit using this language, to illustrate our approach. The reader familiar with the language of tangles will have no difficulty here, and will notice that we have

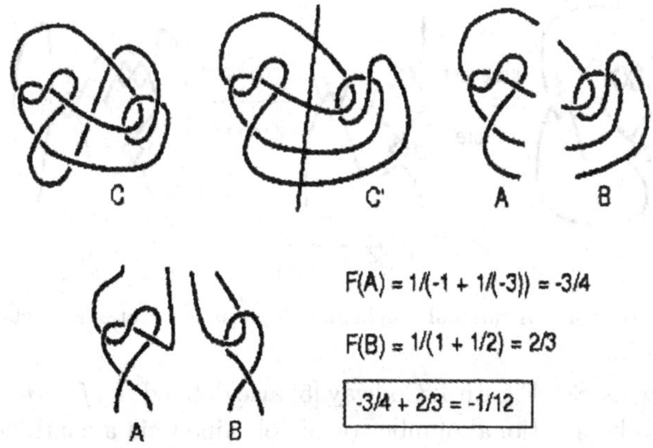

Figure 3. Cutting the culprit into rational tangles.

decomposed the culprit as the numberator of the sum of two tangles whose fractions are $-3/4$ and $2/3$. Since $-3/4+2/3 = -1/12$ we see that the sum of the fractions satisfies the conditions of Theorem 5. *Other readers may wish to read the next section and then come back to this discussion.*

2. Rational Tangles, Rational Knots and Continued Fractions

In this section we recall the subject of rational tangles and rational knots and their relationship with the theory of continued fractions. By the term "knots" we will refer to both knots and links, and whenever we really mean "knot" we shall emphasize it. Rational knots and links comprise the simplest class of links. They are also known in the literature as Viergeflechte, four-plats or 2-bridge knots depending on their geometric representation. The notion of a tangle was introduced in 1967 by Conway [5] in his work on enumerating and classifying knots and links.

A 2-*tangle* is a proper embedding of two unoriented arcs and a finite number of circles in a 3-ball B^3, so that the four endpoints lie in the boundary of B^3. A *tangle diagram* is a regular projection of the tangle on an equatorial disc of B^3. By "tangle" we will mean "tangle diagram". A *rational tangle* is a special case of a 2-tangle obtained by applying consecutive twists on neighbouring endpoints of two trivial arcs. Such a pair of arcs comprise the [0] or [∞] tangles (see Figure 5), depending on their position in the plane. We shall say that the rational tangle is in *twist form* when it is obtained by such successive twists. For examples see Figure 8. Conway defined the rational knots as "numerator" or "denominator" closures of the

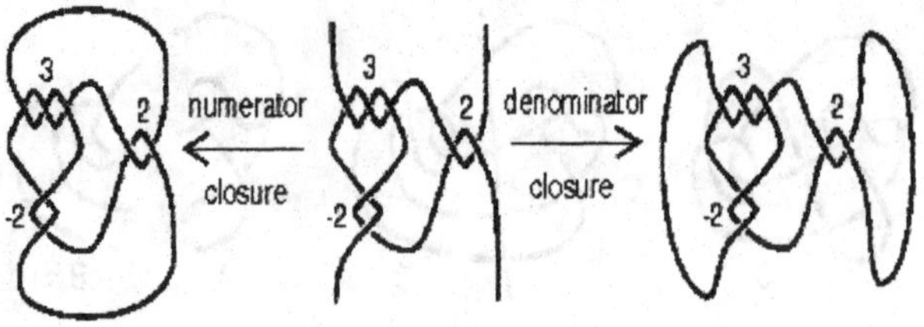

Figure 4. A rational tangle and its closures to rational knots.

rational tangles. See Figure 4. Conway [5] also defined *the fraction* of a rational tangle to be a rational number or ∞, obtained via a continued fraction that is associated with the tangle. We discuss this construction below.

We are interested in tangles up to isotopy. Two tangles, T, S, in B^3 are *isotopic*, denoted by $T \sim S$, if and only if any two diagrams of them have identical configurations of their four endpoints on the boundary of the projection disc, and they differ by a finite sequence of the Reidemeister moves [31, 32], which take place in the interior of the disc. Of course, each twisting operation used in the definition of a rational tangle changes the isotopy class of the tangle to which it is applied. Rational tangles are classified by their fractions by means of the following theorem, different proofs of which are given in [3, 4, 17, 29] and [22].

Theorem 1. (Conway, 1975) *Two rational tangles are isotopic if and only if they have the same fraction.*

More than one rational tangle can yield the same or isotopic rational knots, and the equivalence relation between the rational tangles is mapped into an arithmetic equivalence of their corresponding fractions. Indeed we have:

Theorem 2. (Schubert, 1956) *Suppose that rational tangles with fractions $\frac{p}{q}$ and $\frac{p'}{q'}$ are given (p and q are relatively prime. Similarly for p' and q'.) If $K(\frac{p}{q})$ and $K(\frac{p'}{q'})$ denote the corresponding rational knots obtained by taking numerator closures of these tangles, then $K(\frac{p}{q})$ and $K(\frac{p'}{q'})$ are isotopic if and only if*

1. *$p = p'$ and*
2. *either $q \equiv q'$ mod p or $qq' \equiv 1$ mod p.*

Different proofs of Theorem 2 are given in [3, 23, 34].

2.1. *Rational Tangles and their Invariant Fractions*

We shall now recall from [22] the main properties of rational tangles and of continued fractions, which illuminate the classification of rational tangles. The elementary rational tangles are displayed as either horizontal or vertical twists, and they are enumerated by integers or their inverses, see Figure 5.

The crossing types of 2-tangles (and of unoriented knots) follow the checkerboard rule: shade the regions of the tangle in two colors, starting from the left outside region with grey, and so that adjacent regions have different colors. Crossings in the tangle are said to be of "positive type" if they are arranged with respect to the shading as exemplified in Figure 5 by the tangle $[+1]$, i.e. they have the region on the right shaded as one walks towards the crossing along the over-arc. Crossings of the reverse type are said to be of "negative type' and they are exemplified in Figure 5 by the tangle $[-1]$.

In the class of 2-tangles we have the non-commutative operations *addition* and *multiplication*, as illustrated in Figure 6, which are denoted by "+" and "$*$" respectively. These operations are well-defined up to isotopy. A rational tangle in twist form is created inductively by consecutive additions of the tangles $[\pm 1]$ on the right or on the left and multiplications by

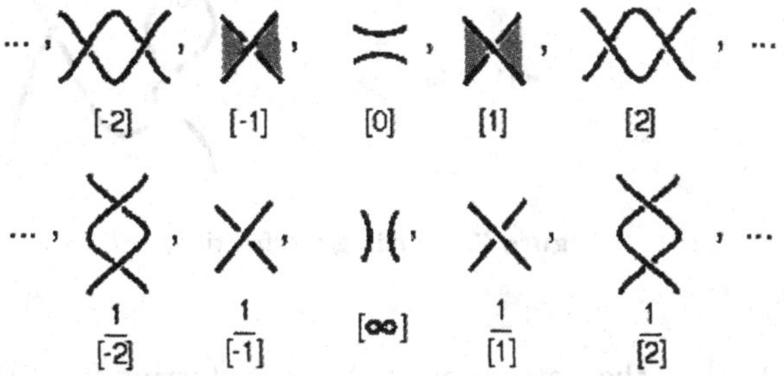

Figure 5. The elementary rational tangles and the types of crossings.

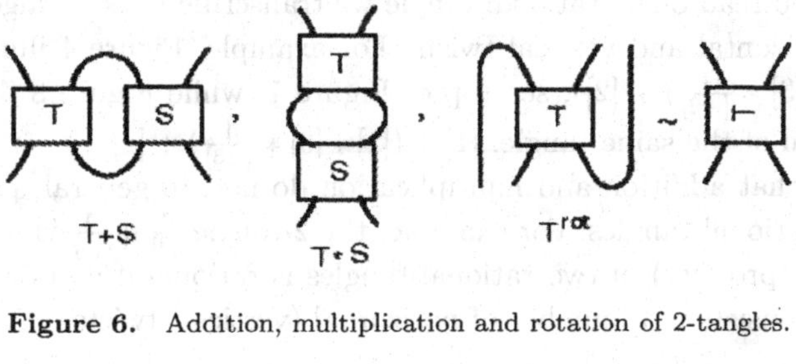

Figure 6. Addition, multiplication and rotation of 2-tangles.

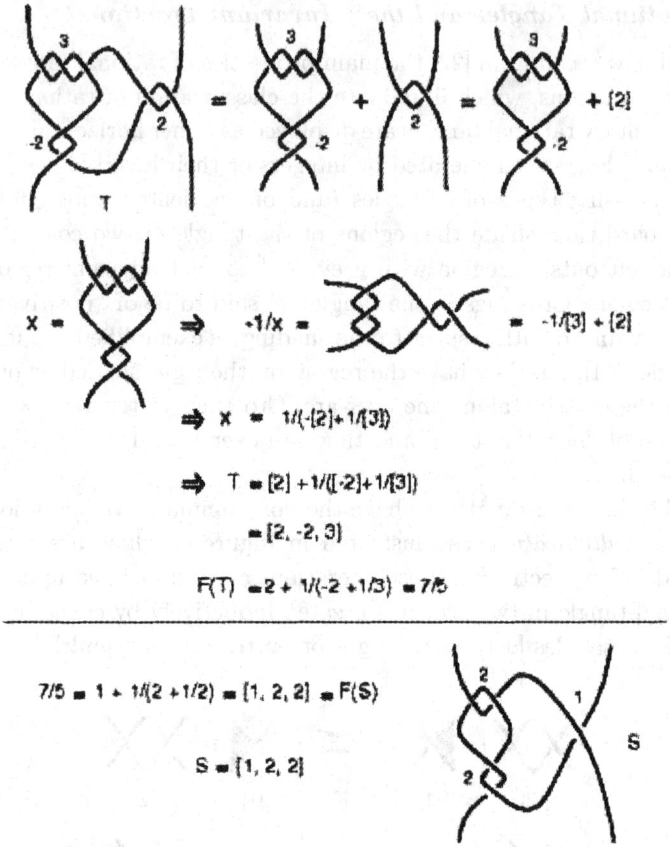

T
+ [2]
x = ⇒ -1/x = = -1/[3] + [2]
⇒ x = 1/(-[2]+1/[3])
⇒ T = [2] + 1/([-2]+1/[3])
= [2, -2, 3]
F(T) = 2 + 1/(-2 + 1/3) = 7/5
7/5 = 1 + 1/(2 + 1/2) = [1, 2, 2] = F(S)
S = [1, 2, 2]
S

Figure 7. Finding the fraction.

the tangles $[\pm 1]$ at the bottom or at the top, starting from the tangle $[0]$ or $[\infty]$. Since the very first crossing can be equally seen as horizontal or vertical, we may always assume that we start twisting from the tangle $[0]$. In order to read out a rational tangle we transcribe it as an algebraic sum using horizontal and vertical twists. For example, Figure 4 illustrates the tangle $(([3] * \frac{1}{[-2]}) + [2])$, see top of Figure 7, while Figure 8 illustrates a twist form of the same tangle: $[1] + ([1] * [3] * \frac{1}{[-3]}) + [1]$.

Note that addition and multiplication do not, in general, preserve the class of rational tangles. For example, the 2-tangle $\frac{1}{[3]} + \frac{1}{[3]}$ is not rational. The sum (product) of two rational tangles is rational if and only if one of the two consists in a number of horizontal (vertical) twists.

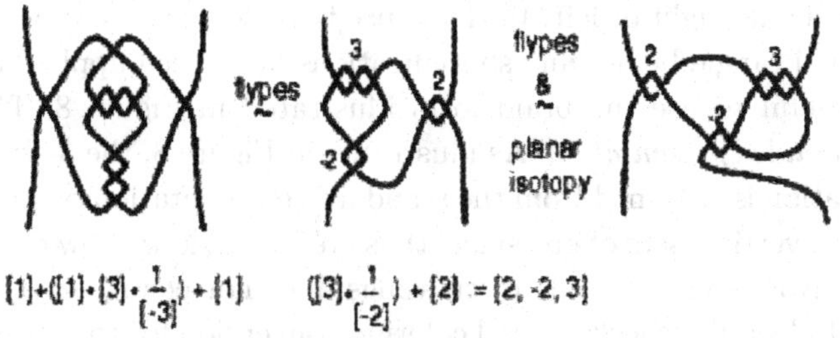

Figure 8. A rational tangle in twist form converted to its standard form and to its 3-strand-braid representation.

The *mirror image* of a tangle T, denoted $-T$, is T with all crossings switched. For example, $-[n] = [-n]$ and $-\frac{1}{[n]} = \frac{1}{[-n]}$. Then, the *subtraction* is defined as $T - S := T + (-S)$. The *rotation* of T, denoted T^{rot}, is obtained by rotating T on its plane counterclockwise by 90°. The *inverse* of T is defined to be $-T^{rot}$. Thus, inversion is accomplished by rotation and mirror image. Note that T^{rot} and the inverse of T are in general not isotopic to T and they are order 4 operations. But for rational tangles the inversion is an operation of order 2 (this follows from the flipping lemma discussed below). For this reason we shall denote the inverse of a rational tangle T by $1/T$, and hence the rotation of the tangle T will be denoted by $-1/T$. This explains the notation for the tangles $\frac{1}{[n]}$.

There is a fraction associated to a rational tangle R which characterizes its isotopy class (Theorem 1). In fact, the fraction is defined for any 2-tangle and always has the following three properties. These suffice for computing the fraction $F(R)$ inductively for rational tangles:

- $F([\pm 1]) = \pm 1$.
- $F(T + S) = F(T) + F(S)$.
- $F(T^{rot}) = -1/F(T)$.

In Figure 7 we illustrate this process by using only these three rules to compute a specific tangle fraction. In the following discussion we discuss the fraction in more detail and how it is related to the continued fraction structure of the rational tangles.

We shall then say that the rational tangle as shown in Figure 8 is in *standard form*. In this Figure we illustrate how to convert a tangle that is in "twist form" to standard form and to the braided form discussed below. Twist form is obtained from two parallel strands by successive twisting at the top, bottom, right or left. In this sense twist form is the general picture of a rational tangle before any simplifications have been applied to it.

It is useful to use the braid form illustrated in Figure 8. This is the 3-*strand-braid representation.* As illustrated in Figure 8, the 3-strand-braid representation is obtained from the standard representation by planar rotations of the vertical sets of crossings, thus creating a lower row of horizontal crossings. Note that the type of crossings does not change by this planar rotation. Indeed the checkerboard coloring convention for the crossing signs identifies the signs as unchanged. Nevertheless, the crossings on the lower row of the braid representation appear to be of opposite sign, since when we rotate them to the vertical position we obtain crossings of the opposite type in the local tangles.

One can associate to a rational tangle in standard form a vector of integers $(a_1, a_2, \ldots, a_n)$, where the first entry denotes the place where the tangle starts untwisting and the last entry where it begins to twist. For example the tangle of Figure 4 corresponds to the vector $(2, -2, 3)$.

Note that the set of twists of a rational tangle may be always assumed *odd.* Indeed, let n be even and let the left-most twist $[a_1]$ be on the upper part of the braid representation. Then, the right-most crossing of the last twist $[a_n]$ may be assumed upper, so that $[a_n]$ can break into $a_n - 1$ lower crossings and one upper. Up to the ambiguity of the right-most crossing, the vector associated to a rational tangle is *unique*, i.e. $(a_1, a_2, \ldots, a_n) = (a_1, a_2, \ldots, a_n - 1, 1)$, if $a_n > 0$, and $(a_1, a_2, \ldots, a_n) = (a_1, a_2, \ldots, a_n + 1, -1)$, if $a_n < 0$. See Figure 9.

Another move that can be applied to a 2-tangle is a *flip*, its rotation in space by 180°. We denote T^{hflip} a horizontal flip (rotation around a horizontal axis on the plane of T) and T^{vflip} a vertical flip. See Figure 10 for illustrations. Note that a flip switches the endpoints of the tangle and, in general, a flipped tangle is not isotopic to the original one. *Rational tangles have the remarkable property that they are isotopic to their horizontal or vertical flips.* We shall refer to this as the *Flipping Lemma.*

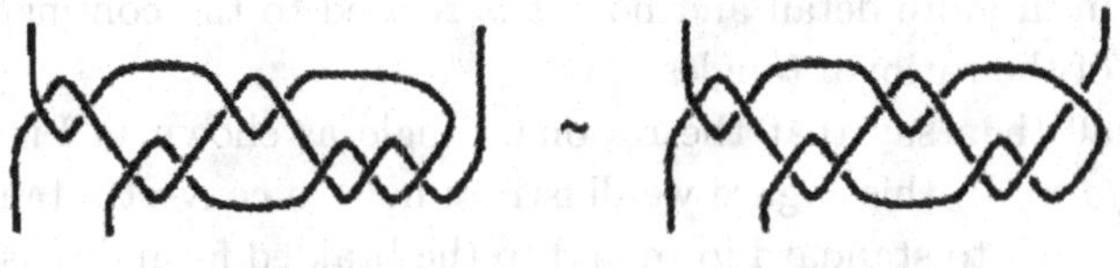

Figure 9. The ambiguity of the first crossing.

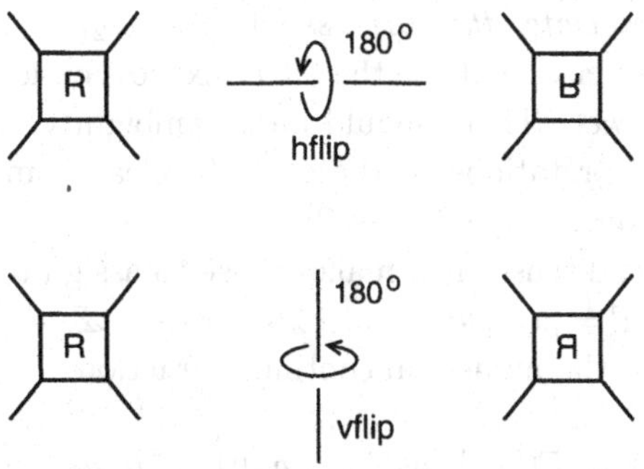

Figure 10. The horizontal and the vertical flip.

A consequence of the Flipping Lemma is that addition and multiplication by $[\pm 1]$ are commutative. Another consequence of the Flipping Lemma is that rotation and inversion of rational tangles each have order two. In particular, rotation is defined via a ninety degree turn of the tangle either to the left or to the right. With this in mind the reader can easily deduce the formula below:

$$T * \frac{1}{[n]} = \frac{1}{[n] + \frac{1}{T}}$$

Indeed, rotate $T * \frac{1}{[n]}$ by ninety degrees and note that it becomes $-[n] - \frac{1}{T}$. Use this to deduce that the original tangle is the negative reciprocal of this tangle. This formula implies that the two operations: addition of $[+1]$ or $[-1]$ and inversion between rational tangles suffice for generating the whole class of rational tangles. As for the fraction, we have the corresponding formula

$$F\left(T * \frac{1}{[n]}\right) = \frac{1}{n + \frac{1}{F(T)}}.$$

The above equation for tangles leads to the fact that a rational tangle in standard form can be described algebraically by a continued fraction built from the integer tangles $[a_1]$, $[a_2]$, $\ldots$, $[a_n]$ with all numerators equal to 1, namely by an expression of the type:

$$[[a_1], [a_2], \ldots, [a_n]] := [a_1] + \frac{1}{[a_2] + \cdots + \frac{1}{[a_{n-1}] + \frac{1}{[a_n]}}}$$

for $a_2, \ldots, a_n \in \mathbb{Z} - \{0\}$ and n even or odd. We allow $[a_1]$ to be the tangle $[0]$. Then, a rational tangle is said to be in *continued fraction form*.

We shall abbreviate the expression $[[a_1], [a_2], \ldots, [a_n]]$ *by writing* $[a_1, a_2, \ldots, a_n]$, and later will use the latter expression for a numerical continued fraction as well. There should be no ambiguity between the tangle and numerical interpretations, as these will be clear from context. Figure 4 illustrates the rational tangle $[2, -2, 3]$.

From the above discussion it makes sense to assign to a rational tangle in standard form, $T = [[a_1], [a_2], \ldots, [a_n]]$, for $a_1 \in \mathbb{Z}$, $a_2, \ldots, a_n \in \mathbb{Z} - \{0\}$ and n even or odd, the numerical continued fraction

$$\begin{aligned} F(T) &= F([[a_1], [a_2], \ldots, [a_n]]) = [a_1, a_2, \ldots, a_n] \\ &:= a_1 + \cfrac{1}{a_2 + \cdots + \frac{1}{a_{n-1} + \frac{1}{a_n}}}, \end{aligned}$$

If a rational tangle T changes by an isotopy, the associated continued fraction form may also change. However, the fraction is a topological invariant of T and does not change. For example, $[2, -2, 3] = [1, 2, 2] = \frac{7}{5}$, see Figure 7. The fraction characterizes the isotopy class of T. For the isotopy type of a rational tangle T with fraction $\frac{p}{q}$ we shall use the notation $[\frac{p}{q}]$. We have omitted here the proof of the invariance of the fraction. The interested reader can consult [5, 17, 22] for various proofs of this fact.

The key to the exact correspondence of fractions and rational tangles lies in the construction of a canonical alternating form for the rational tangle.

We shall say that the rational tangle $S = [\beta_1, \beta_2, \ldots, \beta_m]$ is in *canonical form* if S is alternating and m is odd. From the above, S alternating implies that the β_i's are all of the same sign. It turns out that the canonical form for S is unique. In Figure 11 we bring our working rational tangle $T = [2, -2, 3]$ to its canonical form $S = [1, 2, 2]$. As noted above, $F(T) = F(S) = \frac{7}{5}$.

By Euclid's algorithm and keeping all remainders of the same sign, one can show that every continued fraction $[a_1, a_2, \ldots, a_n]$ can be transformed to a unique canonical form $[\beta_1, \beta_2, \ldots, \beta_m]$, where all β_i's are positive or all negative integers and m is odd. For example, $[2, -2] = [1, 1, 1] = \frac{3}{2}$. There is also an algorithm that can be applied directly to the initial continued

Figure 11. Reducing to alternating form using the swing moves.

fraction to obtain its canonical form, which works in parallel with the algorithm for the canonical form of rational tangles. Indeed, we have:

Proposition 1. *The following identity is true for continued fractions and it is also a topological equivalence of the corresponding tangles:*

$$[\ldots, a, -b, c, d, e, \ldots] = [\ldots, (a-1), 1, (b-1), -c, -d, -e, \ldots].$$

This identity gives a specific inductive procedure for reducing a continued fraction to all positive or all negative terms. In the case of transforming to all negative terms, we can first flip all signs and work with the mirror image. Note also that

$$[\ldots, a, b, 0, c, d, e, \ldots] = [\ldots, a, b+c, d, e, \ldots]$$

will be used in these reductions.

Proof. The technique for the reduction is based on the formula

$$a + 1/(-b) = (a-1) + 1/(1 + 1/(b-1)).$$

If a and b are positive, this formula allows the reduction of negative terms in a continued fraction. The identity in the Proposition follows immediately from this formula. □

2.2. *Rational Knots and Continued Fractions*

By joining with simple arcs the two upper and the two lower endpoints of a 2-tangle T, we obtain a knot called the *Numerator* of T, denoted by $N(T)$. A rational knot is defined to be the numerator of a rational tangle. Joining with simple arcs each pair of the corresponding top and bottom endpoints of T we obtain the *Denominator* of T, denoted by $D(T)$, see Figure 4. We have $N(T) = D(T^{rot})$ and $D(T) = N(T^{rot})$. As we shall see in the next section, the numerator closure of the sum of two rational tangles is still a rational knot. But the denominator closure of the sum of two rational tangles is not necessarily a rational knot, think for example of the sum $\frac{1}{[3]} + \frac{1}{[3]}$.

Given two different rational tangle types $[\frac{p}{q}]$ and $[\frac{p'}{q'}]$, when do they close to isotopic rational knots? The answer is given in Theorem 2. Schubert classified rational knots by finding canonical forms via representing them as 2-bridge knots. In [23] we give a new combinatorial proof of Theorem 2, by posing the question: given a rational knot diagram, at which places may one cut it so that it opens to a rational tangle? We then pinpoint two distinct categories of cuts that represent the two cases of the arithmetic equivalence of Schubert's theorem. The first case corresponds to the *special cut*, as illustrated in Figure 12. The two tangles $T = [-3]$ and $S = [1] + \frac{1}{[2]}$ are non-isotopic by the Conway Theorem, since $F(T) = -3 = 3/-1$, while $F(S) = 1 + 1/2 = 3/2$. But they have isotopic numerators: $N(T) \sim N(S)$, the left-handed trefoil. Now $-1 \equiv 2\, mod\, 3$, confirming Theorem 2. See [23] for a complete analysis of the special cut.

The second case of Schubert's equivalence corresponds to the *palindrome cut*, an example of which is illustrated in Figure 13. Here we see that the tangles

$$T = [2,3,4] = [2] + \frac{1}{[3] + \frac{1}{[4]}}$$

and

$$S = [4,3,2] = [4] + \frac{1}{[3] + \frac{1}{[2]}}$$

both have the same numerator closure. Their corresponding fractions are

$$F(T) = 2 + \frac{1}{3 + \frac{1}{4}} = \frac{30}{13} \quad \text{and} \quad F(S) = 4 + \frac{1}{3 + \frac{1}{2}} = \frac{30}{7}.$$

Note that $7 \cdot 13 \equiv 1\, mod\, 30$.

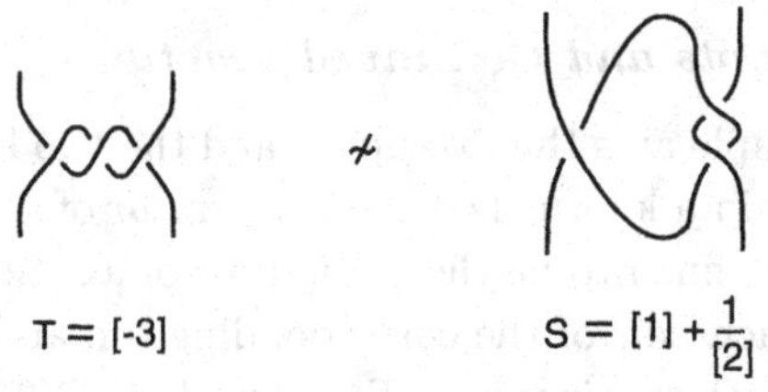

Figure 12. An example of the special cut.

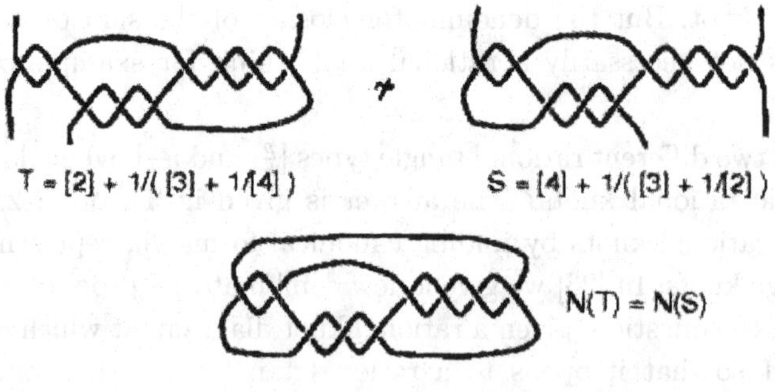

Figure 13. An instance of the palindrome equivalence.

In the general case if $T = [a_1, a_2, \ldots, a_n]$, we shall call the tangle $S = [a_n, a_{n-1}, \ldots, a_1]$ *the palindrome of* T. Clearly these tangles have the same numerator. In order to check the arithmetic in the general case of the palindrome cut we need to generalize this pattern to arbitrary continued fractions and their palindromes (obtained by reversing the order of the terms).

The next Theorem is a known result about continued fractions. See [22, 35] or [24]. We shall omit our proof of this statement. For this we will first present a way of evaluating continued fractions via 2×2 matrices (compare with [15, 26]). This method of evaluation is crucially important for the rest of the paper. We define matrices $M(a)$ by the formula

$$M(a) = \begin{pmatrix} a & 1 \\ 1 & 0 \end{pmatrix}.$$

These matrices $M(a)$ are said to be the *generating matrices* for continued fractions, as we have:

Theorem 3. (The matrix product interpretation for continued fractions) *Let* $\{a_1, a_2, \ldots, a_n\}$ *be a collection of n integers, and let*

$$\frac{P}{Q} = [a_1, a_2, \ldots, a_n]$$

and

$$\frac{P'}{Q'} = [a_n, a_{n-1}, \ldots, a_1].$$

Then $P = P'$ *and* $QQ' \equiv (-1)^{n+1} \bmod P$.

In fact, for any sequence of integers $\{a_1, a_2, \ldots, a_n\}$ *the value of the corresponding continued fraction*

$$\frac{P}{Q} = [a_1, a_2, \ldots, a_n]$$

is given through the following matrix product

$$M = M(a_1)M(a_2)\cdots M(a_n)$$

via the identity

$$M = \begin{pmatrix} P & Q' \\ Q & U \end{pmatrix}$$

where this matrix also gives the evaluation of the palindrome continued fraction

$$[a_n, a_{n-1}, \ldots, a_1] = \frac{P}{Q'}.$$

Proof. We omit the proof of this Theorem. □

3. Sums of Two Rational Tangles

In this section we note that the numerator of the sum of two rational tangles is a rational knot or link. We characterize the knot or link that emerges from this process.

Theorem 4. (Addition of Rational Tangles) *Let* $\{a_1, a_2, \ldots \ldots, a_n\}$ *be a collection of integers, so that*

$$\frac{P}{Q} = [a_1, a_2, \ldots, a_n].$$

Let $\{b_1, b_2, \ldots, b_m\}$ *be another collection of integers, so that*

$$\frac{R}{S} = [b_1, b_2, \ldots, b_m].$$

Let $A = [\frac{P}{Q}]$ *and* $B = [\frac{R}{S}]$ *be the corresponding rational tangles. Then the knot or link* $N(A + B)$ *is rational, and in fact*

$$N(A + B) = N([a_n, a_{n-1}, \ldots, a_2, a_1 + b_1, b_2, \ldots, b_m]).$$

Proof. View Figure 14. In this figure we illustrate a special case of the Theorem. The geometry of reconnection in the general case should be clear from this illustration. □

The next result tells us when we get the unknot.

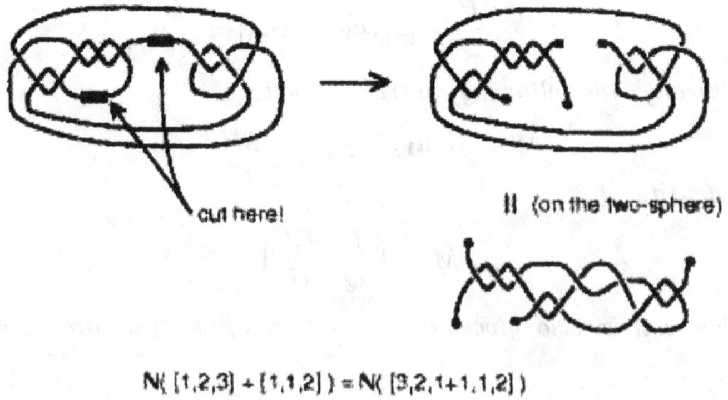

Figure 14. The numerator of a sum of rational tangles is a rational link.

Definition 1. Given continued fractions $\frac{P}{Q} = [a_1, \ldots, a_n]$ and $\frac{R}{S} = [b_1, \ldots, b_m]$, let

$$[a_1, \ldots, a_n] \sharp [b_1, \ldots, b_m] = [a_n, \ldots, a_2, a_1 + b_1, b_2, \ldots, b_m].$$

If

$$\frac{F}{G} = [a_n, \ldots, a_2, a_1 + b_1, b_2, \ldots, b_m],$$

we shall write

$$\frac{P}{Q} \sharp \frac{R}{S} = \frac{F}{G}.$$

Note that $\frac{F}{G}$ is a fraction such that $N([\frac{F}{G}]) = N([\frac{P}{Q}] + [\frac{R}{S}])$.

Theorem 5. *Let*

$$\frac{P}{Q} = [a_1, a_2, \ldots, a_n]$$

and

$$\frac{R}{S} = [b_1, b_2, \ldots, b_m]$$

be as in the previous Theorem. Then

$$N\left(\left[\frac{P}{Q}\right] + \left[\frac{R}{S}\right]\right)$$

is unknotted if and only if $PS + QR = \pm 1$.

Proof. We omit the proof. □

4. Continued Fractions, Convergents and Lots of Unknots

Consider a rational fraction, its corresponding continued fraction, and its matrix representation:

$$P/Q = [a_1, \ldots, a_n]$$

with

$$M = M(\vec{a}) = M(a_1) \cdots M(a_n) = \begin{pmatrix} P & Q' \\ Q & U \end{pmatrix}.$$

Note that since the determinant of this matrix is $(-1)^n$, we have the formula $PU - QQ' = (-1)^n$ from which it follows that

$$P/Q - Q'/U = (-1)^n/QU.$$

Hence, by Theorem 5, the diagram

$$N([P/Q] - [Q'/U])$$

is unknotted and, as we shall see, is a good candidate to produce a hard unknot. Furthermore, the fraction Q'/U has an interpretation as the truncation of our continued fraction $[a_1, \ldots, a_n]$:

$$Q'/U = [a_1, \ldots, a_{n-1}].$$

To see this formula, let

$$N = M(a_1) \cdots M(a_{n-1}) = \begin{pmatrix} R & S' \\ S & V \end{pmatrix},$$

so that

$$R/S = [a_1, \ldots, a_{n-1}].$$

Then

$$\begin{aligned} \begin{pmatrix} P & Q' \\ Q & U \end{pmatrix} &= M(a_1) \cdots M(a_{n-1}) M(a_n) = N M(a_n) \\ &= \begin{pmatrix} R & S' \\ S & V \end{pmatrix} \begin{pmatrix} a_n & 1 \\ 1 & 0 \end{pmatrix} \\ &= \begin{pmatrix} Ra_n + S' & R \\ Sa_n + V & S \end{pmatrix}. \end{aligned}$$

This shows that $Q'/U = R/S = [a_1, \ldots, a_{n-1}]$, as claimed.

Definition 2. One says that $[a_1, \ldots, a_{n-1}]$ is a *convergent* of $[a_1, \ldots, a_{n-1}, a_n]$. We shall say that two fractions P/Q and R/S are *convergents* if the continued fraction of one of them is a convergent of the other.

We see from the above calculation that the two consecutive integers PU and QQ' produce two continued fractions $P/Q = [a_1, \ldots, a_n]$ and $Q'/U = [a_1, \ldots, a_{n-1}]$ so that the second fraction is a convergent of the first.

We have proved the following result.

Theorem 6. *Let P/Q and Q'/U be fractions such that the continued fraction of Q'/U is a convergent of the continued fraction of P/Q. Then*

$$N([P/Q] - [Q'/U])$$

is an unknot.

Proof. The proof is given in the discussion above. □

Remark 1. This Theorem applies to Figure 3, and our early discussion of the Culprit.

The property of one fraction being a convergent of the other is in fact, always a property of fractions produced from consecutive integers. We make this statement formally in the next Lemma (see also [10]).

Lemma 1. *Let P and Q be relatively prime integers and let s and r be a pair of integers such that $Ps - Qr = \pm 1$. Let $R = r + tP$ and $S = s + tQ$ where t is any integer. Then $\{R, S\}$ comprises the set of all solutions to the equation $PS - QR = \pm 1$. If $Ps - Qr = \pm 1$ and $PS - QR = \mp 1$, Then all solutions are given in the form $R = -r + tP$ and $S = -s + tQ$.*

Proof. We omit the proof of this Lemma. □

Theorem 7. *Let P and Q be relatively prime integers and let $P/Q = [a_1, \ldots, a_n]$ be a continued fraction expansion for P/Q. Let $r/s = [a_1, \ldots, a_{n-1}]$ be the convergent for $[a_1, \ldots, a_n]$. Let $R = r + tP$ and $S = s + tQ$ where t is any integer. Then $R/S = [a_1, \ldots, a_n, t]$. Thus P/Q is a convergent of R/S. We conclude that if P/Q and R/S satisfy the condition that $N([P/Q] - [R/S])$ is an unknot, then one of P/Q and R/S is a convergent of the other.*

Proof. We omit the proof of this Theorem. □

5. Constructing Hard Unknots

In this section we indicate how to construct hard unknots by using positive alternating tangles A and B such that $N(A - B)$ is unknotted. By our main results we know how to construct infinitely many such pairs of tangles by taking a continued fraction and its convergent, with the corresponding tangles in reduced (alternating) form.

Let's begin with the case of $5/4 = [1,4] = [1,3,1]$ and $4/3 = [1,3]$. In Figure 15 we show the standard representations of $[1,4]$ and $[1,3]$ as tangles, and the corresponding construction for the diagram of $K = N([1,4] - [1,3])$. The reader will note that this diagram is a hard unknot with 9 crossings, one less than our original Culprit of Figure 3. We give another version of it in Figure 19 (equivalent to its mirror image H). In Section 6 we show that

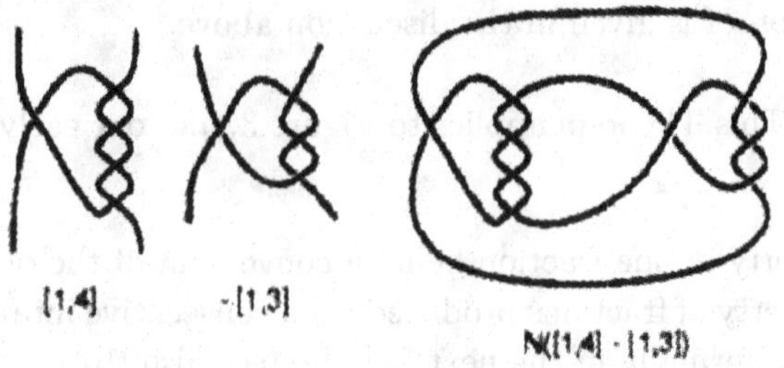

Figure 15. $K = N([1,4] - [1,3])$.

H is one of a small collection of minimal hard unknot diagrams having the form $N(A - B)$ for reduced positive rational tangle diagrams A and B.

In most cases, if one takes the standard representations of the tangles A and B, and forms the diagram for $N(A-B)$, the resulting unknot diagram will be hard. There are some exceptions however, and the next example illustrates this phemomenon.

In Figure 16 we show the standard representations of $[1,3]$ and $[1,2]$ as tangles, and the corresponding construction for the diagram of $N([1,3] - [1,2])$. This diagram, while unknotted, is not a hard unknot diagram due to the three-sided outer region. This outer region allows a type III Reidemeister move on the surface of the two dimensional sphere. In this example, tucking an arc does not create a hard unknot from the given diagram (there is be a type III move available after the tuck).

The Tucking Construct. Figure 17 shows a way to remedy this situation. Here we have replaced $[1,2]$ by $[1,2]^{vflip}$, the 180 degree turn of the tangle $[1,2]$ about the vertical direction in the page. Now we see that the literal diagram of $N([1,3] - [1,2]^{vflip})$ is of course still unknotted and is also not a hard unknot diagram. However this diagram can be converted to an unknot

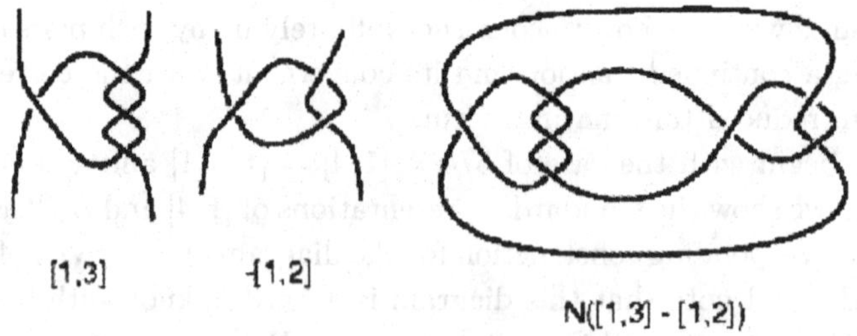

Figure 16. $N([1,3] - [1,2])$.

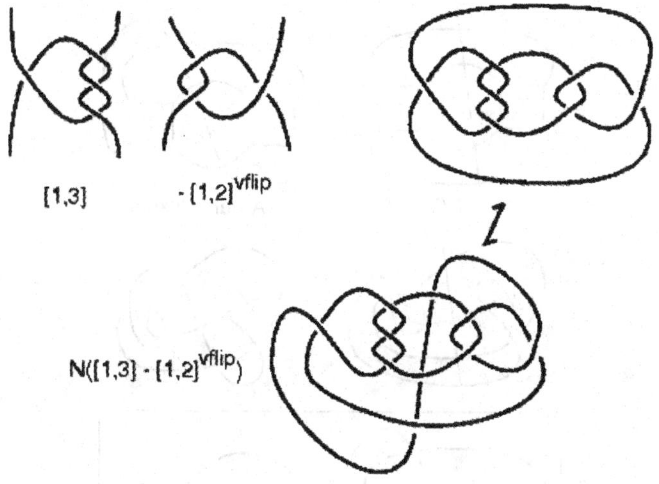

Figure 17. $N([1,3]-[1,2]^{vflip})$.

diagram by tucking an arc as shown in the Figure. The resulting hard unknot is the same diagram of 10 crossings that we had in Figure 3 as our initial Culprit. Note that the other possibility of flipping both tangles in Figure 16 or flipping the first tangle do not lead to a hard unknots. We call this strategem the *tucking construct.* Tucking is accompanied by the vertical flip of on one of the tangles to avoid the placement of a Reidemeister move of type III as a result of the tuck.

The Culprit Revisited. Let's consider the example in Figure 3 again. Here we have $P/Q = F(A) = -3/4$ and $R/S = F(B) = 2/3$. We have $P/Q + R/S = -3/4 + 2/3 = -1/12$. Thus $N([-3/4] + [2/3])$ is an unknot by Theorem 5. This is exactly the unknot C' illustrated in Figure 3.

We can make infinitely many examples of this type. View Figure 18. The pattern is as follows. Suppose that $T = [P/Q]$ and $T' = [R/S]$ are rational tangles such that $PS - QR = \pm 1$. Then we know that $N(T - (T')^{vflip})$ is an unknot. Furthermore we can assume that each of the tangles T and T' are in alternating form. The two tangle fractions have opposite sign and hence the alternation of the weaves in each tangle will be of opposite type. We create a new diagram for $N(T - (T')^{vflip})$ by putting an arc from the bottom of the closure entirely underneath the diagram as shown in Figure 18. This is an example of a sucessful tucking construct. Note how in the example shown in Figure 18, the knot diagram resulting from the tucking construction is indeed our original hard unknot diagram. There

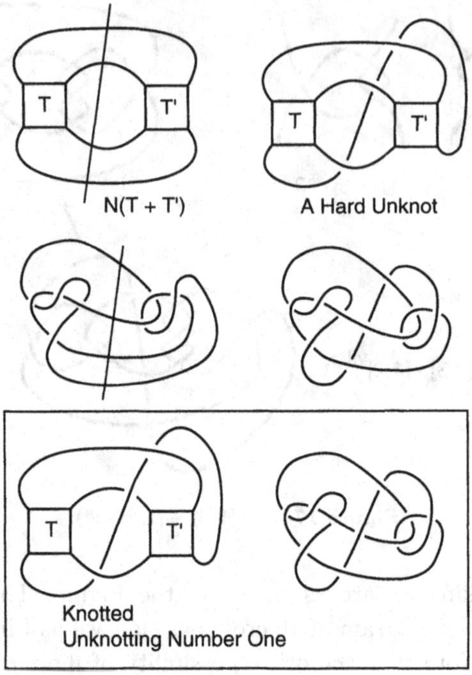

Figure 18. The tucking construct.

are no simplifying Reidemeister moves and there are no moves of type III available on the diagram.

6. The Smallest Hard Unknots

Figure 19 illustrates two hard unknot diagrams H and J with 9 crossings.

Conjecture 1. *Up to mirror images and flyping tangles in the diagrams, the hard unknot diagrams H and J of 9 crossings, shown in Figure 19 ($K = -H$ appears earlier in Figure 15), have the least number of crossings among all hard unknot diagrams.*

Two equivalent versions of the diagram H appear in Figure 19. The right-hand version of H in this figure is of the form

$$H = N([1 + 1/3] - [1 + 1/4]) = N([1, 3] - [1, 4]) = N([4/3] - [5/4]).$$

Note that $[1, 3]$ and $[1, 4] = [1, 3, 1]$ are convergents. Note also that the diagram K of Figure 15 is given by $K = N([1, 4] - [1, 3]) = -N([1, 3] - [1, 4]) = -H$. Thus H and K are mirror images of each other.

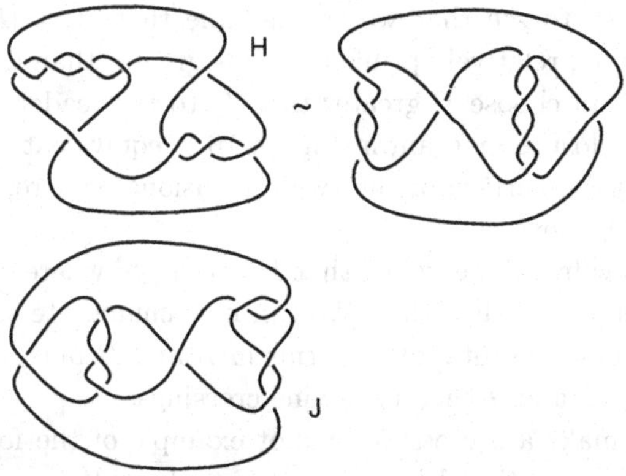

Figure 19. *H* and *J* are hard unknots of 9 crossings.

The diagram J in Figure 19 is of the form

$$N([1+1/3]-[1+1/(2+1/2)]) = N([1,3]-[1,2,2]) = N([4/3]-[7/5]).$$

Note that $[1,3] = [1,2,1]$ and $[1,2,2] = [1,2,1,1]$ are convergents.

Note also that the crossings in J corresponding to 1 in $[1,3]$ and -1 in $-[1,2,2] = [-1,-2,-2]$ can be switched and we will obtain another diagram J', arising as sum of two alternating rational tangles, that is also a hard unknot. This diagram can be obtained from the diagram J without switching crossings by performing flypes (A flype is a turn of a tangle by π that carries a crossing to the other side of the tangle.) on the subtangles $[1,3]$ and $[1,2,2]$ of J, and then doing an isotopy of this new diagram on the two dimensional sphere. (We leave the verification of this statement to the reader.) Thus the diagram J' can be obtained from J by flyping. A similar remark applies to the diagram H, giving a corresponding diagram H', but in this case H' is easily seen to be equivalent to H by an isotopy that does not involve any Reidemeister moves. Thus, up to these sorts of modifications, we have produced essentially two hard diagrams with 9 crossings. Other related hard unknot diagrams of 9 crossings can be obtained from these by taking mirror images.

We have the following result.

Theorem 8. *The diagrams H and J shown in Figure* 19 *are, up to flyping subtangle diagrams and taking mirror images, the smallest hard unknot diagrams in the form $N(A-B)$ where A and B are rational tangles in reduced positive alternating form.*

Proof. It is easy to see that we can assume that $A = [P/Q]$ where P and Q are positive, relatively prime and P is greater than Q. We leave the proof that one can choose P greater than Q to the reader, with the hint: Verify that the closure diagram in Figure 16 is equivalent to the diagram in Figure 19 on the surface of the two dimensional sphere, without using any Reidemeister moves.

We then know from Theorem 8 that $B = [-R/S]$ where one of P/Q and R/S is a convergent of the other. We can now enumerate small continued fractions. We know the total of all terms in A and B must be less than or equal to 9 since H and J each have nine crossings.

In order to make a 9 crossing unknot example of the form $N(A - B)$ where A and B are rational tangles in reduced positive alternating form, we must partition the number 9 into two parts corresponding to the number of crossings in each tangle. It is not hard to see that we need to use the partition $9 = 4 + 5$ in order to make a hard unknot of this form. Furthermore, 4 must correspond to the the continued fraction $[1, 3]$, as $[2, 2]$ will not produce a hard unknot when combined with another tangle. Thus, for producing 9 crossing examples we must take $A = [1, 3]$. Then, in order that A and B be convergents, and B have 5 crossings, the only possibilities for B are $B = [1, 4]$ and $B = [1, 2, 2]$. These choices produce the diagrams H, H', J, J'. It is easy to see that no diagrams with less than 9 crossings will suffice to produce hard unknots, due to the appearance of Reidemeister moves related to the smaller partitions. This completes the proof. □

7. The Goeritz Unknot

The earliest appearance of a hard unknot is a 1934 paper of Goeritz [16]. In this paper Goeritz gives the hard unknot shown in Figure 20. As the reader can see (for example by twisting vertically the tangle $[-3]$ twice), this example is a variant on $N([4] + [-3])$ which is certainly unknotted. The Goeritz example G has 11 crossings, due to the extra two twists that make it a hard unknot. It is part of an infinite family based on $N([n] + [-n + 1])$.

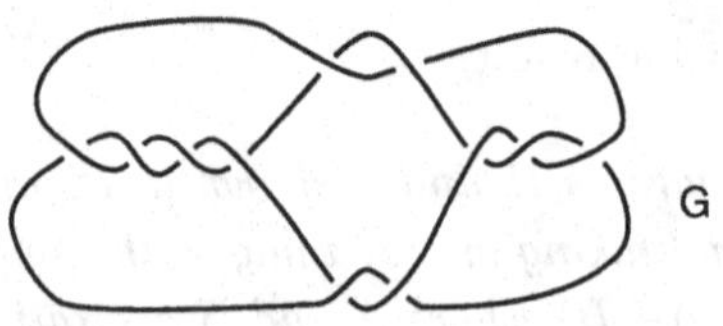

Figure 20. The Goeritz hard unknot.

8. Stability in Processive DNA Recombination

In this section we use the techniques of this paper to study properties of processive DNA recombination topology. Here we use the tangle model of DNA recombination [13, 14, 36] developed by C. Ernst and D.W. Sumners. In this model the DNA is divided into two regions corresponding to two tangles O and I and a recombination site that is associated with I. This division is a model of how the enzyme that performs the recombination traps a part of the DNA, thereby effectively dividing it into the tangles O and I. The recombination site is represented by another tangle R. The entire arrangement is then a knot or link $K[R] = N(O + I + R)$. We then consider a single recombination in the form of starting with $R = [0]$, the zero tangle, and replacing R with the tangle $[1]$ or the tangle $[-1]$. Processive recombination consists in consecutively replacing again and again by $[1]$ or by $[-1]$ at the same site. Thus, in processive recombination we obtain the knots and links

$$K[n] = N(O + I + [n]).$$

The knot or link $K[0] = N(O + I)$ is called the *DNA substrate*, and the tangle $O + I$ is called the *substrate tangle*. It is of interest to obtain a uniform formula for knots and links $K[n]$ that result from the processive recombination.

In some cases the substrate tangle is quite simple and is represented as a single tangle $S = O+I$. For example, we illustrate processive recombination in Figure 21 with $S = [-1/3] = [0, -3]$ and $I = [0]$ with $n = 0, 1, 2, 3, 4$. Note that by Proposition 1 of Section 2.1,

$$\begin{aligned} K[n] &= N(S + [n]) = N([0, -3] + [n]) = N([-3, 0 + n]) = N([-3, n]) \\ &= N(-[3, -n]) = N(-[2, 1, n - 1]). \end{aligned}$$

This formula gives the abstract form of all the knots and links that arise from this recombination process. We say that the formula

$$K[n] = N(-[2, 1, n - 1])$$

for $n > 1$ is *stabilized* in the sense that all the terms in the continued fraction have the same sign and the n is in one single place in the fraction. In general, a *stabilized fraction* will have the form

$$N(\pm[a_1, a_2, \ldots a_{k-1}, a_k + n, a_{k+1}, \ldots, a_n])$$

where all the terms a_i are positive for $i \neq k$ and a_k is non-negative.

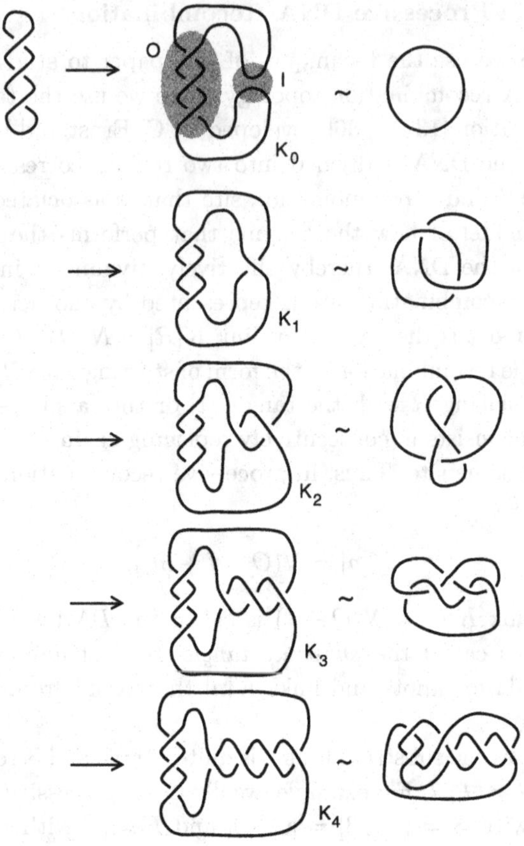

Figure 21. Processive recombination with $S = [-1/3]$.

Let's see what the form of the processive recombination is for an arbitrary sequence of recombinations. We start with

$$O = [a_1, a_2, \ldots, a_{r-1}, a_r]$$
$$I = [b_1, b_2, \ldots, b_{s-1}, b_s].$$

Then

$$\begin{aligned} K[n] &= N(O + (I + [n])) \\ &= N([a_1, a_2, \ldots, a_{r-1}, a_r] + [n + b_1, b_2, \ldots, b_{s-1}, b_s]) \\ K[n] &= N([a_r, a_{r-1}, \ldots, a_2, a_1 + n + b_1, b_2, \ldots, b_{s-1}, b_s]). \end{aligned}$$

Proposition 2. *The formula*

$$K[n] = N([a_r, a_{r-1}, \ldots, a_2, a_1 + n + b_1, b_2, \ldots, b_{s-1}, b_s])$$

can be simplified to yield a stable formula for the processive recombination when n is sufficiently large.

Proof. Apply Proposition 1 of Section 2.1. □

Here is an example. Suppose we take $O = [1, 1, 1, 1]$ and $I = [-1, -1, -1]$ so that the DNA substrate is an (Fibonacci) unknot. (I is the negative of the convergent of O.) Then by the above calculation

$$K[n] = N([1, 1, 1, 1 + n + (-1), -1, -1]) = N([1, 1, 1, n, -1, -1]).$$

Suppose that n is positive. Applying the reduction formula of Proposition 1, we get

$$\begin{aligned} K[n] &= N([1, 1, 1, n, -1, -1]) \\ &= N([1, 1, 1, n - 1, 1, 0, 1]) = N([1, 1, 1, n - 1, 2]), \end{aligned}$$

and this is a stabilized form for the processive recombination.

More generally, suppose that $O = [a_1, a_2, \ldots, a_n]$ where all of the a_i are positive. Let $I = [-a_1, -a_2, \ldots, -a_{n-1}]$. Then $K[0] = N(O + I)$ is an unknotted substrate by our result about convergents. Consider $K[n]$ for positive n. We have

$$\begin{aligned} K[n] &= N([a_n, a_{n-1}, \ldots, a_2, a_1 + n - a_1, -a_2, \ldots, -a_{n-1}]) \\ &= N([a_n, a_{n-1}, \ldots, a_2, n, -a_2, \ldots, -a_{n-1}]) \\ &= N([a_n, a_{n-1}, \ldots, a_2, (n - 1), 1, (a_2 - 1), a_3, \ldots, a_{n-1}]). \end{aligned}$$

If $a_2 - 1$ is not zero, the process terminates immediately. Otherwise there is one more step. In this way the knots and links proceeding from the recombination process all have a uniform stabilized form. Further successive recombination just adds more twist in one entry in the continued fraction diagram whose closure is $K[n]$.

The reader may be interested in watching a visual demonstration of these properties of DNA recombination. For this, we recommend the program *Ginterface* (TangleSolver) [37] of Mariel Vasquez. Her program can be downloaded from the internet as a Java applet, and it performs and displays DNA recombination. Figure 22 illustrates the form of display for this program. The reader should be warned that the program uses the reverse order from our convention when listing the terms in a continued fraction. Thus we say $[1, 2, 3, 4]$ while the program uses $[4, 3, 2, 1]$ for the same structure.

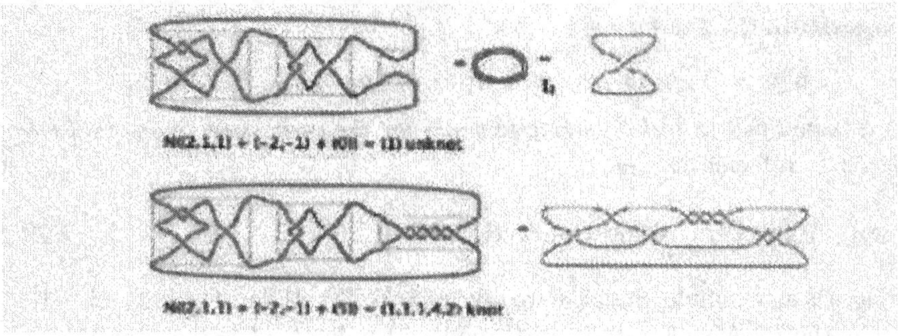

Figure 22. Processive recombination with $S = [1,1,1,1] + [-1,-1,-1]$.

Acknowledgements

The first author thanks the National Science Foundation for support of this research under NSF Grant DMS-0245588. It gives both authors pleasure to acknowledge the hospitality of the Mathematisches Forschungsinstitut Oberwolfach, the University of Illinois at Chicago and the National Technical University of Athens, Greece, where much of this research was conducted. We particularly thank Slavik Jablan for conversations and for helping us, with his computer program LinKnot, to find some key omissions in our initial enumerations.

References

[1] M. Aigner and G. M. Ziegler, "Proofs from The Book", Springer-Verlag (1991).

[2] M. Bruckheimer and A. Arcavi, Farey series and Pick's area theorem, *The Mathematical Intelligencer*, **17**, No. 2 (1995) 64–67.

[3] G. Burde, Verschlingungsinvarianten von Knoten und Verkettungen mit zwei Brücken, *Math. Zeitschrift*, **145** (1975) 235–242.

[4] G. Burde and H. Zieschang, "Knots", de Gruyter Studies in Mathematics **5** (1985).

[5] J. H. Conway, An enumeration of knots and links and some of their algebraic properties, *Proceedings of the conference on Computational problems in Abstract Algebra held at Oxford in 1967*, J. Leech ed., (First edition 1970), Pergamon Press, 329–358.

[6] J. H. Conway and R. K. Guy, "The Book of Numbers", Springer Verlag (1996).

[7] I. A. Dynnikov, Arc presentations of links: Monotonic simplification, ArXiv:Math.GT/0208153 v2 8 Sept. 2003.

[8] H. M. Edwards, "Riemann's Zeta Function", Academic Press (1974), Dover Publications (2001).

[9] J. Farey, On a curious property of vulgar fractions, *Philosophical Magazine*, **47**, (1816) 385–386.

[10] L. R. Ford, Fractions, *American Mathematical Monthly*, **45**, No. 9 (Nov., 1938) 586–601

[11] J. Franel and E. Landau, Les suites de Farey et le problème des nombres premiers, *Gottinger Nachr.*, (1924) 198–206.

[12] D. Epstein and C. Gunn, "Supplement to Not Knot", (booklet accompanying the video "Not Knot"),<http://www.geom.uiuc.edu/video/NotKnot/>, A. K. Peters, Ltd., Natick, MA 01760–4626, USA.

[13] C. Ernst and D. W. Sumners, A calculus for rational tangles: Applications to DNA Recombination, *Math. Proc. Camb. Phil. Soc.*, **108** (1990) 489–515.

[14] C. Ernst and D. W. Sumners, Solving tangle equations arising in a DNA recombination model. *Math. Proc. Cambridge Philos. Soc.*, **126**, No. 1 (1999) 23–36.

[15] J. S. Frame, Continued fractions and matrices, Classroom notes, C. B. Allendoerfer ed., *The Amer. Math. Monthly*, **56** (1949) 98–103.

[16] L. Goeritz, Bemerkungen zur Knotentheorie, *Abh. Math. Sem. Univ. Hamburg*, **10** (1934) 201–210.

[17] J. R. Goldman and L. H. Kauffman, Rational Tangles, *Advances in Applied Math.*, **18** (1997) 300–332.

[18] J. H. Hass and C. Lagarias, The number of Reidemeister moves needed for unknotting, *J. Amer. Math. Soc.*, **14**, No. 2 (2001) 399–428.

[19] G. Hemion, On the classification of the homeomorphisms of 2-manifolds and the classification of three-manifolds, *Acta Math.*, **142**, no. 1–2, (1979) 123–155.

[20] V. F. R. Jones, A polynomial invariant for links via von Neumann algebras, *Bull. Amer. Math. Soc.*, **129** (1985) 103–112.

[21] L. H. Kauffman, State Models and the Jones Polynomial, *Topology*, 26 (1987) 395–407.

[22] L. H. Kauffman and S. Lambropoulou, On the classification of rational tangles, *Advances in Applied Math*, **33**, No. 2 (2004) 199–237.

[23] L. H. Kauffman and S. Lambropoulou, On the classification of rational knots. *L'Enseignement Mathématiques*, **49** (2003) 357–410.

[24] A. Kawauchi, "A Survey of Knot Theory", Birkhäuser Verlag (1996).

[25] A. Ya. Khinchin, "Continued Fractions", Dover (1997) (republication of the 1964 edition of Chicago Univ. Press).

[26] K. Kolden, Continued fractions and linear substitutions, *Archiv for Math. og Naturvidenskab*, **6** (1949) 141–196.

[27] K. Millett (private conversation at Institutes Hautes Etudes Scientifiques, Burres Sur Yevette, France, circa 1988).

[28] H. Morton, An irreducible 4-string braid with unknotted closure, *Proc. Camb. Phil. Soc.*, **93** (1983) no. 2, 259–261.

[29] J. M. Montesinos, Revetêments ramifiés des noeuds, Espaces fibres de Seifert et scindements de Heegaard, *Publicaciones del Seminario Mathematico Garcia de Galdeano, Serie II, Seccion 3* (1984).

[30] C. D. Olds, "Continued Fractions", New Mathematical Library, Math. Assoc. of Amerika, **9** (1963).
[31] K. Reidemeister, "Knotentheorie" (Reprint), Chelsea, New York (1948).
[32] K. Reidemeister, Knoten und Verkettungen, *Math. Zeitschrift*, **29** (1929) 713–729.
[33] R. Scharein, KnotPlot (program available from the web), <http://wren.pims.math.ca/knotplot>.
[34] H. Schubert, Knoten mit zwei Brücken, *Math. Zeitschrift*, **65** (1956) 133–170.
[35] L. Siebenmann, Lecture Notes on Rational Tangles, Orsay (1972) (unpublished).
[36] D. W. Sumners, Untangling DNA, *Math.Intelligencer*, **12** (1990) 71–80.
[37] M. Vasquez, TangleSolver (program available from the web),<http://math.berkeley.edu/mariel/>.
[38] H. S. Wall, "Analytic Theory of Continued Fractions", D. Van Nostrand Company, Inc. (1948).

DEVELOPING A MATHEMATICAL MODEL OF PHAGOCYTOSIS: A LEARNING PROCESS

NATAŠA MACURA

Department of Mathematics,
Trinity University,
One Trinity Place, San Antonio, TX 78212

TONG ZHANG and ARTURO CASADEVALL

Albert Einstein College of Medicine,
1300 Morris Park Ave., Bronx, NY 10461

August 13, 2006

We describe a basic mathematical model of phagocytosis. We outline the ideas involved in building the model as well as possible directions for further developments and refinements of the current model.

1. Introduction

Our mathematical exploration of phagocytosis originated in our interest in the mechanisms of antibody-mediated protection against *C. neoformans* and the relationship between antibody dose and protective efficacy. Passive immunization with antibody to the capsule is protective but the dose-response shows the peculiar finding that administration of large amounts of antibody abrogates protection [4, 6], and such doses can actually enhance the course of infection. This phenomenon has been called a "prozone-like" effect. While studying the interaction of macrophages and *C. neoformans* in vitro we noted that the efficacy of phagocytosis declines at higher antibody concentrations [4, 6]. Given that this observation could be associated with the prozone-like phenomenon observed in passive protection experiments we decided to study in more detail and construct a mathematical model of phagocytosis that would allow us to better understand the contribution of the various parameters to opsonic efficacy.

Phagocytosis is a fundamental mechanism of host defense whereby macrophages and neutrophils ingest and destroy microbial cells. The process of phagocytosis follows a complex choreography whereby the microbe first attaches to the host cells it is subsequently ingested through various receptor signaling paths that are not well understood. There have been several attempts to generate mathematical models of phagocytosis in the literature [2, 7]. However, none of the models available have addressed the underlying biophysical mechanisms of phagocytosis and the critical contribution of opsonin concentration. Moreover, the subject was last investigated almost two decades ago.

Our goal is to use the available knowledge about phagocytosis to construct a mathematical model that will provide a framework for the further studies and rigorous description of the process and compare predictions from the model with experimental data. We initially drafted a set of differential equations derived from common assumptions but found that it did not account for the experimental observations. Consequently, we carried out additional experimental work and model adjustments that in turn provided new insights into this fundamentally important host defense mechanism.

In this paper we describe the main ideas in modeling phagocytosis and pose several questions that need to be addressed in a combination of experimental work and mathematical modeling to understand and formally describe this system.

2. Efficacy of Phagocytosis

Phagocytosis is a process that consists of the engulfment and destruction of extracellularly-derived materials by phagocytic cells, such as macrophages and neutrophils. A large number of experimental studies are described in the literature with a variety of particles and cells, such as erythrocytes, bacteria or yeast cells. Our modeling work is mostly based on experimental studies of phagocytosis of *Cryptococcus neoformans* by macrophages but our overall goal is to construct a general mathematical model that will apply to a variety of organisms and systems and facilitate understanding of common determinants of phagocytosis in different systems as well as enable us to study the differences.

We are particularly interested in determining the variables that influence the *efficacy of phagocytosis*. Efficacy of phagocytosis can be characterized by the number or percentage of cells ingested over a certain period of time and therefore is a variable that is by its nature quantitative. The efficacy of phagocytosis can also be described as the percentage of phagocytic cells that ingest microbes or particles involved. In general, the efficacy of phagocytosis will depend on a variety of factors that differ from one biological model to another, but some of the main determinants like the presence or the absence of antibody, the type and number of receptors involved and the number of cells previously ingested or attached are common to a large class of systems studied experimentally.

We are interested in antibody meditated phagocytosis and a model of efficacy of phagocytosis with respect to the amount of antibody present. We are especially interested in the the dose-response effects of antibody overload on the efficacy of the phagocytic process as well as in its role in the immune response in general.

We find that the most suitable measure of efficacy of phagocytosis for modeling is the number of microbes ingested per phagocyte. This approach enables us to use ideas from population models as well as modeling biochemical reactions.

The system studied is a complex biological system that is not fully understood and requires a combination of phenomenological and biophysical modeling. Modeling such a process requires a series of successive refinements and improvement of the model where each stage is characterized by clear biophysical foundation and is also flexible enough to allow for explorations of alternative hypothesis and adjustments of the model. In this process we start with a phenomenological model based on biochemical reactions models and analyze dependence of the model outcomes on the constants involved. That modeling step identifies new variables and constraints that are then explored experimentally in an effort to both test and refine the model and provide directions for further theoretical and experimental work.

3. Modeling phagocytosis

The earliest and one of a very few models of phagocytosis is developed in [2] where the authors model phagocytosis as a bimolecular reaction and propose the differential equation

$$\frac{dY_I}{dt} = -k(CP_0 - Y_I)(Y_0 - Y_I) \qquad \text{(M-1)}$$

where Y_I is the total number of microbes ingested at time t, Y_0 is the initial number of microbes, P_0 the initial number of phagocytes, and C the phagocytic capacity of a one phagocytic cell. In this model the rate k is constant, which results in a differential equation that is easily solved for the function that describes the growing population of ingested microbes. We remark that a comparison with experimental data in [2] also led to introducing a number of modified models that fit the data better then this basic model with modifications and adjustments for possible variations in the capacity for phagocytosis.

Experimental results in [4–6] confirm that the population of ingested microbes grows exponentially with respect to time, but also reveal that, in the case of antibody mediated phagocytosis, the total number of microbes ingested depends on the initial concentration of free antibody. That is an expected observation since antibody facilitates phagocytosis, see Figure 1, Panel A. However there is also an unexpected phenomenon, that is, that

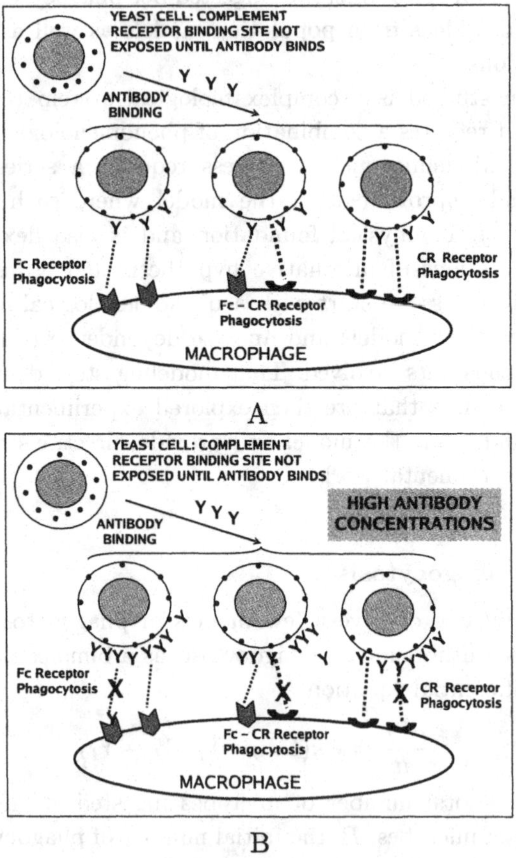

Figure 1. Antibody (IgG) binding to *Cryptococcus neoformans* capsule facilitates phagocytosis through both Fc and complement receptors (Panel A). Too much antibody bound to the capsule might block CR binding sites on the capsule as well as Fc receptors on the macrophage. (Panel B).

higher concentrations of antibody result in a reduction in the efficacy of phagocytosis. This paradoxical dose-response effect has been called a "prozone-like effect."

In our model we test several possible explanations for this phenomenon and design experiments that lead to better understanding of phagocytosis. In [3] we constructed a mathematical model of phagocytosis that describes the dependence of the phagocytic efficacy on the initial concentration of free antibody. Our model is based on a differential equation similar to the equation M-1, that is

$$\frac{dY_I}{dt} = -k_{phago}P_0(Y_0 - Y_I) \tag{M-2}$$

but our efforts are concentrated on understanding how the "constant" k_{phago} depends on the amount of antibody added to our in-vitro phagocytic system. We built our model on the assumption that k_{phago} is a not a constant but a function of both the amount of antibody bound to microbes, Ab, as well as the number R of available receptors. Both above mentioned quantities depend on the initial concentration of free antibody: higher initial concentration of free antibody will result in more antibody bound to the yeast capsule over time and therefore higher rate of phagocytosis. However, at the same time free antibody will also bind to Fc receptors and block them from interacting with antibody bound to the yeast cell, possibly slowing down or even preventing phagocytosis (see Figure 1, Panel B). Our experiments described in Figure 2 confirm that the presence of irrelevant antibody indeed reduces the efficacy of phagocytosis.

The first step in building the model was to determine the main variables and parameters of the system and test our understanding of experimental conditions and their relationship to the process modeled. The process of constructing a model of phagocytosis revealed a number of experimental variables and conditions that could potentially influence the efficiency of phagocytosis and that needed to be incorporated into the modeling process.

One example of such an experimental variable is the importance of the procedure of mixing microbes and macrophages and, in particular, its timing. The importance of this step became apparent when we realized that the dynamics of the binding of antibody capsule, and not just the equilibrium values, might strongly influence the outcome of this experiment since the saturation time was comparable to the time of our experiments.

The identification of such variables led to additional experiments and modification of the initial models.

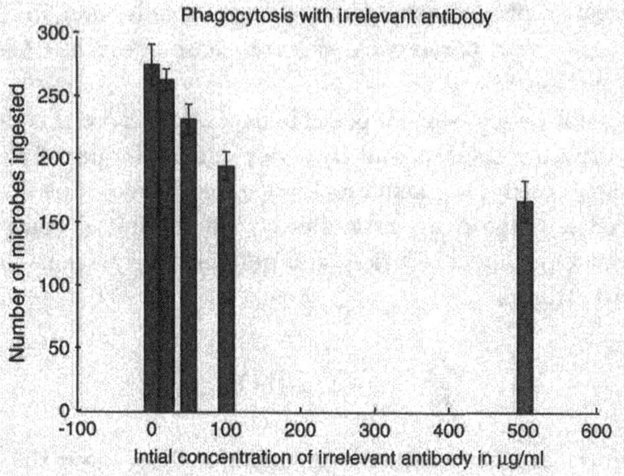

Figure 2. Adding irrelevant antibody, that is antibody that does not bind to the yeast capsule but does bind to Fc receptors, reduces the efficacy of phagocytosis. The concentration of (relevant) IgG in all the experiments was 10 μg/ml while the concentration of irrelevant antibody was varied. The number of ingested microbes was counted after two hours. The number of ingested yeast cells is given per 100 macrophages and the error bars denote the standard deviation.

4. Modeling Goals

The long term goal of our project is to achieve a detailed mathematical description for this basic biological process and therefore to explain biophysical mechanism and biochemical networks that are underlying the process of phagocytosis. This challenging goal will necessarily require sequential refinements of the model that incorporate results from both experimental and theoretical studies of particular aspect of this complex process. We are building a model of a process that is not phenomenologically fully understood and whose understanding could immensely benefit from organizing the existing knowledge and data within a framework of mathematical modeling.

In our basic model our goal is to attempt to understand how the efficacy of phagocytosis depends on the amount of antibody present as well as availability of receptors. We use our model to test a number of different functions of Ab and R as a possible choice of for k_{phago}. The model indicates that the dependence is not linear, which is consistent with the fact that we are modeling a biological system with complex interactions and regulatory mechanisms.

A number of questions will need to be answered in order to explain the dependence of k_{phago} on the the amount of antibody bound to the yeast capsule and the number of receptors and provide the next step in the model refinement. We are currently conducting experiments that will contribute to understanding of the following issues that might affect the efficacy of phagocytosis:

(i) how does the number of receptors influence the rate of phagocytosis,
(ii) is there a minimal surface density of receptors for attachment or ingestion process to start and proceed,
(iii) understanding the rate at which receptors are internalized in phagocytosis,
(iv) the destiny of ingested receptors (recycling or degradation of receptors) as well as synthesis of new receptors during phagocytosis,
(v) contribution of complement receptors to phagocytosis,
(vi) model of interactions and cooperation between receptors of the same type as well as receptors of different types.

Some of these questions are present and unanswered in the literature, and building a mathematical model enables us to connect them and organize our knowledge into a more complete picture of phagocytosis.

In all our current experiments we use the same number of macrophages. Since the multiplication rate for macrophages is rather low, we can assume that the number is constant throughout the experiment. Moreover, since we are using the 1:1 ratio of phagocytes and microbes we can consider, according to published experimental results, that the number of ingested microbes is well below the phagocytic capacity of macrophages even when the loss of macrophages is taken into account. This makes the exponential growth model M-2 very suitable for our experimental conditions.

However, at larger numbers of microbes we expect the phagocytic capacity of phagocytes to limit the total number of ingested microbes. A more advanced mathematical model will not only have to take this capacity into account but can also be used to determine when the capacity of phagocytosis is determined by the actual inherent phagocytic capacity of macrophages or when it is limited by the number of receptors involved and the rate of their synthesis and trafficking to the surface.

Moreover, the phagocytic capacity of macrophages is a characteristic that is probably dependent on the rate of phagocytosis and the health of macrophages. Consequently, this parameter requires careful and controlled experiments when introduced in modeling.

A more detailed mathematical model will also need to address the question of the difference in the mechanisms of attachment and ingestion in phagocytosis. In our basic model we still do not distinguish among the attachment and ingestion, and results in [1] indicate that different mechanisms govern these two steps in phagocytosis: their experimental studies show that while the total number of attached and ingested microbes in this system depends on the amount of antibody, the fraction of attached microbes that also are ingested is constant and does not depend on the amount of antibody. We predict that in a more general setting involving different types of phagocytic receptors, the two processes will differ considerably and a more detailed mathematical model will need to account for the differences.

In our initial model presented in [3] we do not study the signaling cascades that govern the macrophage response, nor the cytoskeletal movement including a possible receptor aggregation and cooperation. These questions can be and will need to be addressed in a more advanced model as well.

The most limiting factor in this type of a study is that we are dealing with a biological system. The system itself is complex and tweaking experimental conditions is not always easy. For example, observing phagocytosis for longer then two hours is experimentally challenging since microbes may impair the health and functioning of phagocytic cells.

A carefully constructed mathematical model is a powerful tool for exploration of such a complex process and enables us to overcome some of the experimental limitations. It also provides us with guidelines for designing and preforming optimal experiments as well combining partial results into a complete and precise description of the process studied.

References

[1] Camner, P., Lundborg, M., Lastbom, L., Gerde, P., Gross, N., Jarstrand, C. Experimental and calculated parameters on particles phagocytosis by alveolar macrophages, *J. Appl. Physiol.* **92**(2002), 2608–16.

[2] Petri, A. S. I., Egerer, R., Suss., J., Schutz, H. Development of mathematical models for an in-vitro phagocytosis test system, *Zentralbl. Bakteriol. Mikrobiol. Hyg. [A]* **267**(1987), 217–27.

[3] Macura, N., Zhang, T., Casadevall, A. A mathematical model of phagocytosis, submitted (2006).

[4] Taborda, C. P., Casadevall, A. Immunoglobulin M efficacy against cryptococcus neoformans: Mechanism, dose-dependence and prozone-like effects in passive protection experiments, *J. Immunol.* **166**(2001), 2100–7.

[5] Taborda, C. P., Casadevall, A. CR3 (CD11b/CD18) and CR4 (CD11c/CD18) are involved in complement-independent antibody-mediated phagocytosis of cryptococcus neoformans, *Immunity* **16**(2002), 791–802.

[6] Taborda, C. P., Rivera, J., Zaragoza, O., Casadevall, A. More is not necessarily better: prozone-like effects in passive immunization with IgG, *J. Immunol.* **170**(2003), 3621–30.

[7] Tran, C. L., Jones, A. D., Donaldson, K. Mathematical model of phagocytosis and inflammation after the inhalation of quartz at different concentrations, *Scand. J. Work Environ. Health* **21** Suppl 2 (1995), 50–54.

AN AGE STRUCTURED MODEL OF T CELL POPULATIONS

BRYNJA KOHLER

Department of Mathematics and Statistics
Utah State University
3900 Old Main Hill
Logan, UT 84321

Although it is well established that T cell populations dramatically fluctuate when the immune system responds to many diseases, many specifics of how T cell activation occurs are not understood. We develop a model that tracks T cell interactions with dendritic cells, which is a crucial process for activating adaptive immune responses. We assume that the duration of stimulation by dendritic cells in the lymph node determines the extent of T cell differentiation. In this paper, we first describe the general partial differential equation model with justifications for the modeling choices. Second, we present a few highlights of the mathematical analysis of the model with certain simplifying assumptions. Finally, we discuss some simulations of the full model and interpret the biological predictions the model suggests.

1. Biological Background

The adaptive immune response to a pathogen typically is initiated through cellular interactions that take place in the lymph node. In a lymph node, a specific naive T cell may contact antigen presented on a mature dendritic cell's major histocompatibility complex (MHC) surface proteins. If the contact is sufficient, this activates the naive T cell, and leads to differentiation and proliferation of that clonal line of cells specific for the antigen:MHC complex.

The dynamics of the T cell population responses are well studied for certain pathogens such as murine Lymphocytic Choreomeningitis Virus. Experiments show a consistent T cell response which consists of an expansion phase in which T cell numbers of a specific lineage rapidly increase several orders of magnitude, a retraction phase in which T cell numbers dramatically drop, and a memory phase in which T cells numbers remain relatively constant at about 5–10 % of the numbers at the peak response. Simple mathematical models have been used to capture these dynamics and estimate parameters such as cell proliferation rates and apoptosis rates. While total populations are well tracked, there is still debate in the literature about the nature of the T cell differentiation process.

Lanzavecchia and Sallusto (2000) propose that T cell differentiation is driven by the duration of T cell receptor (TCR) stimulation and cytokine action, and that hierarchical levels of differentiation are achieved throughout the period of stimulation by dendritic cells [5]. In particular, they believe that memory T cells are generated early after shorter durations of interaction with antigen, while effector cells are generated only from lengthier antigen interaction intervals. This is in contrast to other models which suggest that effector cells arise before memory cells in the differentiation scheme.

Recently, research groups have greatly improved upon our ability to study these dendritic cell–T cell interactions by developing the technology to visualize the cell interactions in preparations of intact lymph nodes using two-photon microscopy ([1] and [7]). This research benefits from clearly laid out mathematical models which describe and track these lymph node interactions. Significant modeling contributions regarding how ligand-receptor binding affects signaling outcomes are reviewed in [2].

In this paper, we develop a model that tracks dendritic cell–T cell interactions, and assuming that T cell differentiation is a function of time spent engaged in a synapse with antigen we create a framework to study differentiation models further. We concentrate on T cell stimulation in the lymph nodes (see Figure 1), since this is where the adaptive immune responses to primary infections are initiated. The goal of the model presented here is to be able to determine the distribution of cells that exit the lymph node as a function of the duration of the TCR stimulation they experience and their antigen avidity, beginning with a single naive cell. We can interpret the resulting population of cells as having different functions that depend on their degree of differentiation, and thus further investigate whether memory cells are intermediates or terminally differentiated T cells. Additionally, we can use our model to investigate competition and selection of T cell clones with differing affinities for antigen.

In the next section, we describe the general model with justifications for our modeling choices. We present some of our mathematical analysis such as the model derivation and a much simplified ODE model that simply defines the net growth or decay of T cells. We also provide exact solutions of the model with certain simplifying assumptions. In the final section, we

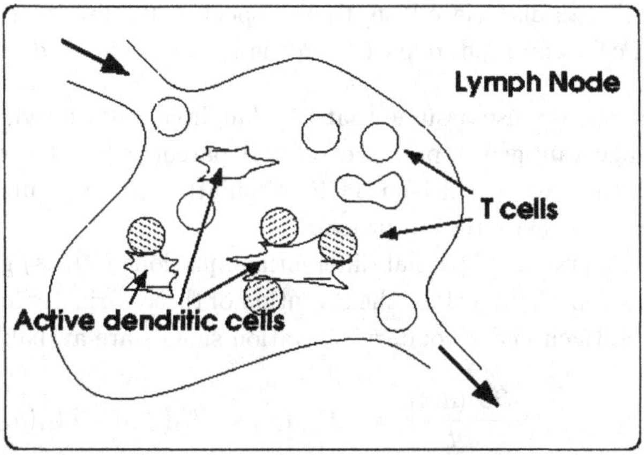

Figure 1. T cell activation in a lymph node. This simplified cartoon illustrates our model of T cell–dendritic cell interactions in a lymph node. A naive T cell that binds to antigen presentation sites on active dendritic cells may become activated, and subsequently differentiate to a proliferating type. The population of cells of that clonal line may continue binding to the antigen presented or leave the lymph node. The mathematical model presented here tracks the bound cells (shaded) and the unbound cells, while keeping track of the total time spent bound to antigen on active dendritic cell for each T cell.

show our results of numerical simulations on the full model and interpret the model's biological predictions.

2. The Model and Analysis

We wish to track T cells not only in time (t), but also in some measure of maturation. Since T cells mature based on the duration of TCR stimulation, we define an additional independent variable, age (a), as the total duration of TCR stimulation experienced by the cell. This variable represents the cumulative stimulation regardless of the length of individual contacts.

We define two state variables for each lineage of T cells to describe the system: $B_i(a,t)$ is the density of T cells from lineage i bound to antigen sites (meaning sites on a dendritic cell where antigen:MHC complexes may bind with T cell receptors) with age a at time t, and $U_i(a,t)$ is the density of unbound T cells in lineage i with age a at time t. We assume that the underlying binding process to dendritic cells displaying antigen is stochastic.

We hypothesize the following rules for the model:

- T cells age (experience TCR stimulation) only while they are bound.

- Bound cells can dissociate from their respective binding sites.
- Unbound cells can bind, depart from the lymph node, or divide.

Importantly, we also assume that the daughter cells of dividing T cells retain the age (antigen experience) of the parent cell. This reflects our assumption that antigen driven TCR stimulation drives a unidirectional hierarchical cell differentiation process.

Hence, the system of partial differential equations (PDEs) given below constitutes a general model for the evolution of these variables in the lymph node when antigen and secondary activation signals are available:

$$\frac{\partial U_i(a,t)}{\partial t} = \lambda_i B_i(a,t) - \alpha U_i(a,t) - kU_i(a,t) + pU_i(a,t) \tag{1}$$

$$\frac{\partial B_i(a,t)}{\partial a} + \frac{\partial B_i(a,t)}{\partial t} = -\lambda_i B_i(a,t) + \alpha U_i(a,t). \tag{2}$$

These equations are derived from the underlying stochastic model in Section 2.1.

We must now more clearly define the transition rates in these equations. We use λ_i for the detachment rate (or unbinding rate) of T cell lineage i, while α represents the attachment (or binding) rate. These rates relate to the overall avidity of the TCR to the binding site, the availability of binding sites, and possibly also to the maturity of the dendritic cell or T cell involved in the interaction. An explicit dependency on age would follow from experimental work that shows that when the two cells interact, key cell surface proteins concentrate to form more efficient synapses [3]. We may assume that the encounter rate α is independent of the cell lineage, and only λ depends on the TCR's avidity for antigen. For simplicity, we will assume that λ_i is constant for a given T cell lineage. This is reasonable because we assume the dissociation rate is a property of the intrinsic conformation of the TCR and its affinity for the antigen peptide presented on MHC. We analyze the system without considering age dependence of α or λ, but leave that for future study.

In our most general formulation, we model α with an explicit dependence on total available binding sites, allowing us to study the competition effects during activation when antigen sites are limited. We let

$$\alpha = \widehat{\alpha}/S\left(S - \int_{-\infty}^{\infty}\sum_i B_i da\right), \tag{3}$$

where $\widehat{\alpha}$ is some maximum binding rate when all binding sites S are available. Then $\widehat{\alpha}/S$ is the binding rate per site, and $(S - \int_{-\infty}^{\infty} \sum_i B_i da)$ is the number of available binding sites. Note that the total number of bound T cells, or the integral of B_i over age, can never exceed S. We further assume that the number of binding sites remains constant for the period of activation. Alternatively, S itself may be a function of time to account for an influx of mature dendritic cells from the infection site or the half–life of the antigen:MHC complex on the dendritic cell surface.

Unbound cells are free to exit the lymph node at a rate k and can divide at a rate p (proliferation rate), regardless of their affinity for antigen presented. We assume that these rates do not depend on the particular clone, and hence, have omitted any subscript i. Physiologically, proliferation depends chiefly on IL-2 production and concentration. Activated cells secrete IL-2, and the locally produced IL-2 is required for cells to divide. If we assume that activated cells produce enough IL-2 to proliferate at their maximum, we can model proliferation with a simple step function. We let

$$p(a) = \widehat{p}H(a - a_p). \tag{4}$$

where $H(a)$ denotes a Heaviside step function, a_p is the age of significant activation so that proliferation can occur, and $\widehat{p}$ is the maximum proliferation rate. Hence, we make p an age dependent function, assuming that IL-2 is available and well–mixed in the lymph node. Any cell that has reached this level of maturity is assumed to be a proliferating cell.

As for the departure rate, k, the simplest thing is to assume k is constant and identical for all clones of unbound cells. More realistically, however, there may be some delay before T cells that have entered the lymph node can depart from it. This could be due to the fact that the cells enter one way (through high endothelial venules), then remain for a while before they can exit into circulation. This would best be modeled with a delay or an additional compartment in the equations. It is also possible that the departure rate is density dependent, implying a lymph node has a limited carrying capacity (N). This could be modeled with a term such as $k = \frac{\widehat{k}N}{|N - \int \sum_i (U_i + B_i) da|}$. However, we prefer the assumption that the removal rate k is constant.

After a certain level of maturity, T cells die due to activation induced cell death (AICD). We could model this process with an additional term in both equations where $d = \widehat{d}H(a - a_d)$ is some large death rate in both the U_i and B_i equations. Again, H is a Heaviside function, and a_d is the critical age at which cells begin to die off rapidly. More simply, we could instead solve the equations on the domain $0 \leq a \leq a_d$ and assume both $U_i(a,t)$ and $B_i(a,t)$ are zero beyond that domain. We are most concerned with deriving the age distribution of cells resulting from antigen priming in the lymph node so this death rate is not so important; we can derive the age distribution and subsequently decide which cells are alive or dead.

Ultimately, we seek to track all the T cells that leave the lymph node and enter the circulation, which we identify as $T_i(a,t)$. They grow according to $\frac{dT_i(a,t)}{dt} = kU_i(a,t)$. Hence, we use the integral of $U_i(a,t)$ over time as a measure of the age distribution of total cells produced of the clonal line with index i.

Initial conditions for this problem may vary depending on what we choose to simulate. We will generally consider the problem of how a single naive clone activates and proliferates. When we consider only one T cell clone we naturally drop the subscript i. We assume this initially unbound naive cell has age 0, implying the initial population mass distribution

$$U(a,0) = \delta(a) \tag{5}$$

where δ is the Dirac delta function. We can further assume we initially have no bound cells, i.e.,

$$B(a,0) = 0. \tag{6}$$

The dynamics apply only in the quadrant where $a > 0$ and $t > 0$. We must translate these very singular initial conditions to the correct boundary conditions for the equivalent Cauchy problem (see Section 2.4). For our numerical simulations in Section 3, we use a discrete version of the Dirac delta function.

We have introduced our model, which consists of equations 1–6. Analysis and special solutions are given in the following section, while we return to the full model and biological interpretations in 3.

2.1. *Derivation of the age-structured PDE system*

We assume a cell can be in either of two states: bound or unbound. A cell's age a is equal to the cumulative time spent bound to antigen on an antigen presenting cell. $U(a,t)$ is defined as the population density of unbound cells of age a at time t. $B(a,t)$ is the density of bound cells of age a at time t. Hence, $\int_{a_1}^{a_2} U(a,t)+B(a,t)da$ is interpreted as the cell population with ages between a_1 and a_2 at time t. We understand $\int_{a_1}^{a_2} U(a,t)da$ to be the number of those that are in the unbound state.

We assume cells can change their state, divide, or depart from the lymph node, and assign the following probabilities to those events:

($\lambda\Delta t$) – probability that a bound cell dissociates in time Δt as $\Delta t \longrightarrow 0$;
($\alpha\Delta t$) – probability that an unbound cell binds in time Δt as $\Delta t \longrightarrow 0$;
($k\Delta t$) – probability that an unbound cell leaves the lymph node in time Δt as $\Delta t \longrightarrow 0$;
($p\Delta t$) – probability that an unbound cell divides in time Δt as $\Delta t \longrightarrow 0$.

Hence we assume these processes are stochastic with Poisson distributions. The state transitions are illustrated in Figure 2 with the proliferation and departure rates omitted.

Now we can write the master equations which govern the system. The unbound cells at age a and time $t + \Delta t$ come from bound cells that have dissociated, unbound cells that have not bound or departed from the lymph node, and from births of new unbound cells. Hence, we have the equation

$$U(a, t + \Delta t) = \lambda \Delta t B(a, t) + (1 - \alpha\Delta t)(1 - k\Delta t)U(a, t) + p\Delta t U(a, t)$$

Bound cells, on the other hand, age in exact accordance with time. After a time step Δt, bound cells that have not dissociated advance an amount $\Delta a = \Delta t$ in age. They also can come from unbound cells which bind at some arbitrary time in the interval between t and $t + \Delta t$. (Suppose this arbitrary time is $t + \Delta t - \Delta x$ where $0 \leq \Delta x < \Delta a = \Delta t$.) This gives us the master equation for the bound cell density

$$B(a + \Delta a, t + \Delta t) = (1 - \lambda\Delta t)B(a, t) + \alpha\Delta t U(a + \Delta x, t).$$

By rearranging these equations we obtain:

$$\frac{U(a, t + \Delta t) - U(a, t)}{\Delta t} = \lambda B(a, t) - \alpha U(a, t) - kU(a, t) + \alpha k \Delta t U(a, t) + pU(a, t)$$

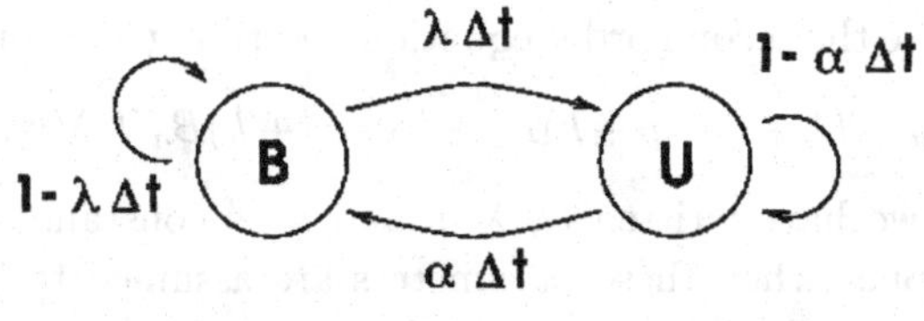

Figure 2. Stochastic transition diagram. This cartoon shows the possible transitions between bound and unbound states, and the transition probabilities.

$$\frac{B(a+\Delta a,t+\Delta t)-B(a,t+\Delta t)}{\Delta a}+\frac{B(a,t+\Delta t)-B(a,t)}{\Delta t}$$
$$=-\lambda B(a,t)+\alpha U(a+\Delta a,t)$$

And in the limit as $\Delta a = \Delta t \longrightarrow 0$ we obtain the linear hyperbolic PDE system:

$$\frac{\partial U(a,t)}{\partial t} = \lambda B(a,t) - \alpha U(a,t) - kU(a,t) + pU(a,t) \quad (7)$$

$$\frac{\partial B(a,t)}{\partial a} + \frac{\partial B(a,t)}{\partial t} = -\lambda B(a,t) + \alpha U(a,t) \quad (8)$$

This model can be written more concisely as:

$$U_t = \lambda B - (\alpha + k - p)U$$
$$B_a + B_t = -\lambda B + \alpha U$$

or in matrix form as:

$$\begin{bmatrix} U \\ B \end{bmatrix}_t + \begin{pmatrix} 0 & 0 \\ 0 & 1 \end{pmatrix} \begin{bmatrix} U \\ B \end{bmatrix}_a = \begin{pmatrix} \gamma & \lambda \\ \alpha & -\lambda \end{pmatrix}$$

with $\gamma = p - (\alpha + k)$.

Letting $v = \begin{bmatrix} U \\ B \end{bmatrix}$, $C = \begin{pmatrix} 0 & 0 \\ 0 & 1 \end{pmatrix}$, and $D = \begin{pmatrix} \gamma & \lambda \\ \alpha & -\lambda \end{pmatrix}$. we have

$$v_t + Cv_a = D.$$

Differentiating the equation 8 with respect to t, we can obtain a single, second order PDE:

$$B_{ta} + B_{tt} = -\lambda B_t + \alpha U_t.$$

Substituting the equation 7 in for U_t then yields:

$$B_{ta} + B_{tt} = -\lambda B_t + \alpha\left[\lambda B - (\alpha - p + k)U\right]$$
$$\Rightarrow B_{ta} + B_{tt} = -\lambda B_t + \alpha\lambda B - (\alpha - p + k)\alpha U$$
$$\Rightarrow B_{ta} + B_{tt} = -\lambda B_t + \alpha\lambda B - (\alpha - p + k)\left[\lambda B + B_a + B_t\right].$$

Finally, we have the second order equation describing the bound population

$$B_{ta} + B_{tt} + (\lambda + \alpha - p + k)B_t + (\alpha - p + k)B_a + \lambda(k - p)B = 0$$

Although we have written α, λ, p and k as constant coefficients, the derivation applies when these parameters are assumed to be functions of time and age as well. This derivation thus gives rise to our most generally formulated model in Section 2.

2.2. *Age independent ODE problem*

Assume α, λ, p and k are constants. Let us study the age independent problem, by integrating our equations over the age variable:

$$\int_{-\infty}^{\infty}\left(\frac{\partial U}{\partial t}\right)da = -\gamma\int_{-\infty}^{\infty} U da + \lambda\int_{-\infty}^{\infty} B da$$

$$\int_{-\infty}^{\infty}\left(\frac{\partial B}{\partial a}+\frac{\partial B}{\partial t}\right)da = \alpha\int_{-\infty}^{\infty} U da - \lambda\int_{-\infty}^{\infty} B da$$

Then since for any t, $B(a,t)$ is bounded with compact support (with respect to a), we can change variables to $u = \int_{-\infty}^{\infty} U da$ and $b = \int_{-\infty}^{\infty} B da$, and look at the ODE system

$$\frac{du}{dt} = -\gamma u + \lambda b \tag{9}$$

$$\frac{db}{dt} = \alpha u - \lambda b. \tag{10}$$

And our initial conditions of one naive unbound cell become:

$$u(0) = 1 \tag{11}$$

$$b(0) = 0. \tag{12}$$

This linear problem is easy to solve!

Note first, that the total cell population evolves according to

$$\frac{d(u+b)}{dt} = (\alpha-\gamma)(u+b), \quad \text{with } (u+b)(0) = 1. \tag{13}$$

So, when $\alpha > \gamma$ the net population grows exponentially, when $\alpha = \gamma$ the population stays fixed ($= 1$) and when $\alpha < \gamma$ the population decays.

The solution of the full system yields a sum of exponentials where the growth or decay depends on the eigenvalues of the matrix $D = \begin{pmatrix} -\gamma & \lambda \\ \alpha & -\lambda \end{pmatrix}$. The eigenvalues are always real when λ, α, and γ are positive real numbers. Eigenvalues are

$$\mu_{1,2} = \frac{1}{2}\left[-(\lambda+\gamma) \pm \sqrt{(\lambda+\gamma)^2 - 4\lambda(\gamma-\alpha)}\right]$$

with respective eigenvectors

$$v_{1,2} = \begin{bmatrix} 1 \\ \dfrac{\alpha}{\mu_{1,2}+\lambda} \end{bmatrix} = \begin{bmatrix} 1 \\ \dfrac{2\alpha}{(\lambda-\gamma) \pm \sqrt{(\lambda+\gamma)^2 - 4\lambda(\gamma-\alpha)}} \end{bmatrix}$$

And hence the exact solution with our initial values is

$$u = -\frac{\mu_1 + \lambda}{\mu_2 - \mu_1}\exp(\mu_1 t) + \frac{\mu_2 + \lambda}{\mu_2 - \mu_1}\exp(\mu_2 t) \tag{14}$$

$$b = \frac{\alpha}{\mu_2 - \mu_1}(\exp(\mu_1 t) + \exp(\mu_2 t)) \tag{15}$$

The eigenvalues are real and distinct provided that

$$(\lambda + \gamma)^2 > 4\lambda(\gamma - \alpha).$$

Provided our parameters λ and α are positive, and γ is real, then we always have real eigenvalues because

$$(\lambda - \gamma)^2 \geq -4\lambda\alpha \Rightarrow (\lambda + \gamma)^2 \geq 4\lambda(\gamma - \alpha).$$

If $\gamma < \alpha$ we have one positive (μ_1) and one negative (μ_2) eigenvalue. The population grows in the "direction" of the eigenvector associated with the positive eigenvalue and decays in the "direction" associated with the negative eigenvalue.

If $\alpha < \gamma$ we have two negative eigenvalues. The population decays to zero as time proceeds.

If $\alpha = \gamma$ we have one zero and one negative eigenvalue. The population will go to an equilibrium value. In fact, for long time, $u \to \frac{\lambda}{\alpha+\lambda}$ and $b \to \frac{\alpha}{\alpha+\lambda}$. In this case the exact solution is

$$u = \frac{\alpha}{\alpha + \lambda}\exp(-(\alpha + \lambda)t) + \frac{\lambda}{\alpha + \lambda} \tag{16}$$

$$b = -\frac{\alpha}{\alpha + \lambda}\exp(-(\alpha + \lambda)t) + \frac{\alpha}{\alpha + \lambda} \tag{17}$$

This analysis confirms that T cell populations grow in the lymph node provided that the proliferation rate, p, is greater than the rate of departure, k.

2.3. *Simplified case: No net population change*

Let us consider the special case where γ is equal to α. This case is relevant biologically if we assume cells do not leave or proliferate in the lymph node. Or, if we consider one binding site and one cell and we wish to understand how the total time bound evolves as a function of time. Interestingly, the same simplified case would result from the assumption that the rate of proliferation is exactly equal to the rate at which cells depart the lymph node.

In this case, we can think of U and B as true probability distributions of unbound and bound cells over age, respectively. If we imagine starting with a single cell of age zero which has never bound to antigen, the probability of binding will follow a Poisson distribution $u_0(t) = e^{-\alpha t}$. The PDE system applies for $a > 0$. Hence we have the conservation law

$$\int_0^\infty U(a,t) + B(a,t)da + u_0(t) = 1 = u_0(0) \quad \forall t.$$

meaning that the probability a cell is unbound or bound of any age (> 0), plus the probability that a cell has never bound equals one.

If we consider a system where $\gamma \neq -\alpha$ the net population will change as a function of time. We then will have

$$\int_0^\infty U(a,t) + B(a,t)da + u_0(t) = n(t) \quad \forall t.$$

where $n(t)$ is the equation for the total population as a function of time.

In the next section we work out the details of formulating the correct boundary conditions for the Cauchy problem in the simplified case.

2.4. *Formulation of the Cauchy problem*

A PDE with side conditions is well posed if it admits a unique solution for any values assigned to the data — or more precisely, if the solution depends continuously on the data [6]. Hyperbolic problems such as the one we have formulated have this attribute; however, due to the singular nature of our initial condition, we must properly formulate boundary data along manifolds transverse to the characteristics.

We consider the system of equations

$$\frac{\partial U}{\partial t} = -\alpha U + \lambda B \tag{18}$$

$$\frac{\partial B}{\partial a} + \frac{\partial B}{\partial t} = \alpha U - \lambda B \tag{19}$$

for $a > 0$. We also assume a function $u_0(t)$ satisfies

$$\frac{du_0}{dt} = -\alpha u_0, \tag{20}$$

representing the naive cell distribution. Let $u_0(0) = u^*$ be the initial naive cells. Total cells will always equal the initial quantity of naive cells, so our conservation law is:

$$u_0(t) + \int_0^\infty (B(a,t) + U(a,t))da = u^* \tag{21}$$

for any $t > 0$. However, $U(a,t) = 0$ and $B(a,t) = 0$ for $a > t$ so that

$$u_0(t) + \int_0^t (B+U)da = u^*.$$

Note that this implies

$$\frac{d}{dt}\left(\int_0^t (B+U)da\right) = \frac{d}{dt}(u^* - u_0(t)) = -\frac{du_0}{dt}.$$

Also note that the sum of equations 18 and 19 yields

$$\frac{\partial(U+B)}{\partial t} + \frac{\partial B}{\partial a} = 0.$$

Fix $a > 0$ and observe that

$$\frac{d}{dt}\int_0^a (B(\hat{a},t) + U(\hat{a},t))d\hat{a} = -\int_0^a \frac{\partial B(\hat{a},t)}{\partial \hat{a}} d\hat{a} = B(0,t) - B(a,t).$$

If $a > t$ then $B(a,t) = 0$, so that

$$\frac{d}{dt}\int_0^a (B(\hat{a},t) + U(\hat{a},t))d\hat{a} = B(0,t) = -\frac{du_0}{dt}$$
$$\Rightarrow B(0,t) = *\alpha u_0(t). \qquad (22)$$

This specifies a boundary condition at $a = 0$.

The above calculation applies only if a is fixed. If not,

$$\frac{d}{dt}\int_0^{a(t)} (B(\hat{a},t) + U(\hat{a},t))d\hat{a} = \frac{da}{dt}(B(a,t) + U(a,t)) + B(0,t) - B(a,t).$$

Letting $a = t$, ($\frac{da}{dt} = 1$),

$$(B(t,t) + U(t,t)) + B(0,t) - B(t,t) = -\frac{du_0}{dt} \Rightarrow U(t,t) = 0. \qquad (23)$$

This specifies a boundary condition along the characteristic curve $\eta = t = a$.

We can also observe that along the curve $\eta = a$, $\eta = t$,

$$\frac{dB}{d\eta} = \alpha U(\eta,t) - \lambda B(\eta,t) = -\lambda B(\eta = t, t).$$

Furthermore, $B(\eta = 0, t = 0) = \alpha u^*$, so that

$$B(t,t) = \alpha u^* \exp(-\lambda t).$$

It turns out that we do not need to define B at this boundary to solve the problem as this comes from the solution of the problem itself, this curve is itself a characteristic.

Hence, the problem we wish to solve is formulated as follows in characteristic coordinates. We change coordinates from (a, t) to (η, τ) where $a = \eta$ and $t = \eta + \tau$. Then we have the equations

$$\frac{\partial U}{\partial \tau} = -\alpha U(\eta, \tau) + \lambda B(\eta, \tau) \tag{24}$$

$$\frac{\partial B}{\partial \eta} = \alpha U(\eta, \tau) - \lambda B(\eta, \tau) \tag{25}$$

and our boundary conditions are

$$U(\eta, 0) = 0 \tag{26}$$

$$B(0, \tau) = \alpha u^* \exp(-\alpha \tau). \tag{27}$$

2.5. *Solution using Laplace transforms*

To solve the problem above, we use the Laplace transforms in τ:

$$\widehat{B}(\eta, s) = \int_0^\infty e^{-s\tau} B(\eta, \tau) d\tau \tag{28}$$

$$\widehat{U}(\eta, s) = \int_0^\infty e^{-s\tau} U(\eta, \tau) d\tau. \tag{29}$$

Equations 25–27 become

$$s\widehat{U} - U(\eta, 0) = \lambda \widehat{B} - \alpha \widehat{U} \quad \Rightarrow \quad \widehat{U} = \frac{\lambda \widehat{B}}{s + \alpha}$$

$$\frac{\partial \widehat{B}}{\partial \eta} = \alpha \widehat{U} - \lambda \widehat{B}$$

with

$$\widehat{B}(0, s) = \frac{\alpha u^*}{\alpha + s}.$$

So,

$$\widehat{B} = \exp\left[\left(\frac{\alpha\lambda}{s + \alpha} - \lambda\right)\eta\right] \frac{\alpha u^*}{\alpha + s}$$

or by rearranging

$$\widehat{B} = \alpha u^* e^{-\lambda\eta} \frac{1}{\alpha + s} \exp\left(\frac{\alpha\lambda\eta}{s + \alpha}\right) \tag{30}$$

From [8] we have that

$$f(s) = \frac{e^{-a/s}}{s^{n+1}}, \qquad n > -1$$

is the Laplace transform of

$$F(t) = \left(\frac{t}{a}\right)^{n/2} J_n(2\sqrt{at})$$

where J_n is the Bessel function of the first kind of order n.

Hence we invert $\widehat{B}$ to find

$$B(\eta, \tau) = \alpha u^* e^{-\lambda\eta} e^{-\alpha\tau} I_0(2\sqrt{\alpha\lambda\eta\tau}) \tag{31}$$

where I_v is the *modified* Bessel function of the first kind of order v. To find U, we must invert

$$\widehat{U} = \frac{\lambda}{\alpha + s} \alpha u^* e^{-\lambda\eta} \frac{1}{\alpha + s} \exp\left(\frac{\alpha\lambda\eta}{s + \alpha}\right) \tag{32}$$

which gives

$$U(\eta, \tau) = \lambda\alpha u^* e^{-\lambda\eta} e^{-\alpha\tau} \left(\frac{\tau}{\alpha\lambda\eta}\right)^{1/2} I_1(2\sqrt{\alpha\lambda\eta\tau}). \tag{33}$$

So, the solution back in (a, t) coordinates is

$$U(a,t) = \lambda\alpha u^* e^{-\lambda a} e^{-\alpha(t-a)} \sqrt{\frac{(t-a)}{\alpha\lambda a}} I_1(2\sqrt{\alpha\lambda a(t-a)})$$
$$B(a,t) = \alpha u^* e^{-\lambda a} e^{-\alpha(t-a)} I_0(2\sqrt{\alpha\lambda a(t-a)})$$

which applies for $0 < a < t$. Or we can write

$$U(a,t) = \lambda\alpha u^* e^{-\lambda a} e^{-\alpha(t-a)} \sqrt{\frac{(t-a)}{\alpha\lambda a}} \times I_1(2\sqrt{\alpha\lambda a(t-a)}) H(a) H(t-a) \tag{34}$$

$$B(a,t) = \alpha u^* e^{-\lambda a} e^{-\alpha(t-a)} I_0(2\sqrt{\alpha\lambda a(t-a)}) H(a) H(t-a), \tag{35}$$

and this is the solution to our problem. We show plots of these solutions in Figures 3, 4 and 5. We have U and B as functions of time and age, and can study these dynamics for a range of parameter values, in this special case where $p = k$ implying there is no net population change in the lymph node. Solutions are depicted in Figures 3–4.

It is clear that solutions for U and B are shaped like humps that move along the a axis and deform. The amplitude decays while the support grows, and the integral approaches a constant. Indeed this system describes some kind of diffusion process. Note that the peaks of the solutions for U and for B move with a speed of $\frac{\alpha}{\alpha+\lambda}$. This makes sense because this is the fraction

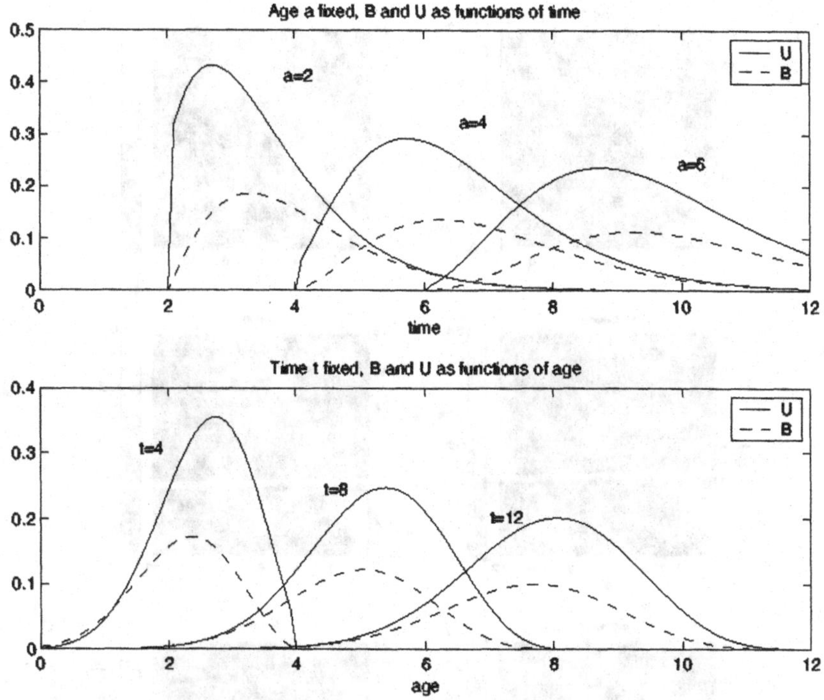

Figure 3. Solution curves with a constant total population. This figure shows solution cross sections when $\lambda = 1$, $\alpha = 2$, and either a is held fixed (top), or t is held fixed (bottom). No population growth is allowed here.

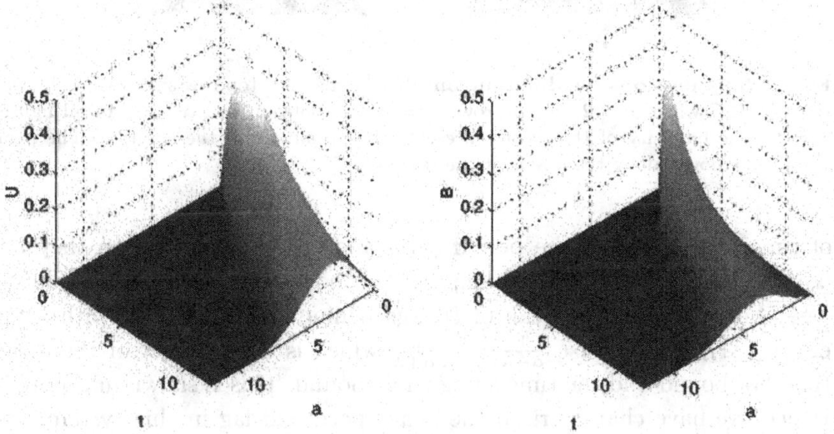

Figure 4. Solution surfaces with a constant total population. Surface plot of solution for $\alpha = .5$ and $\lambda = 1$.

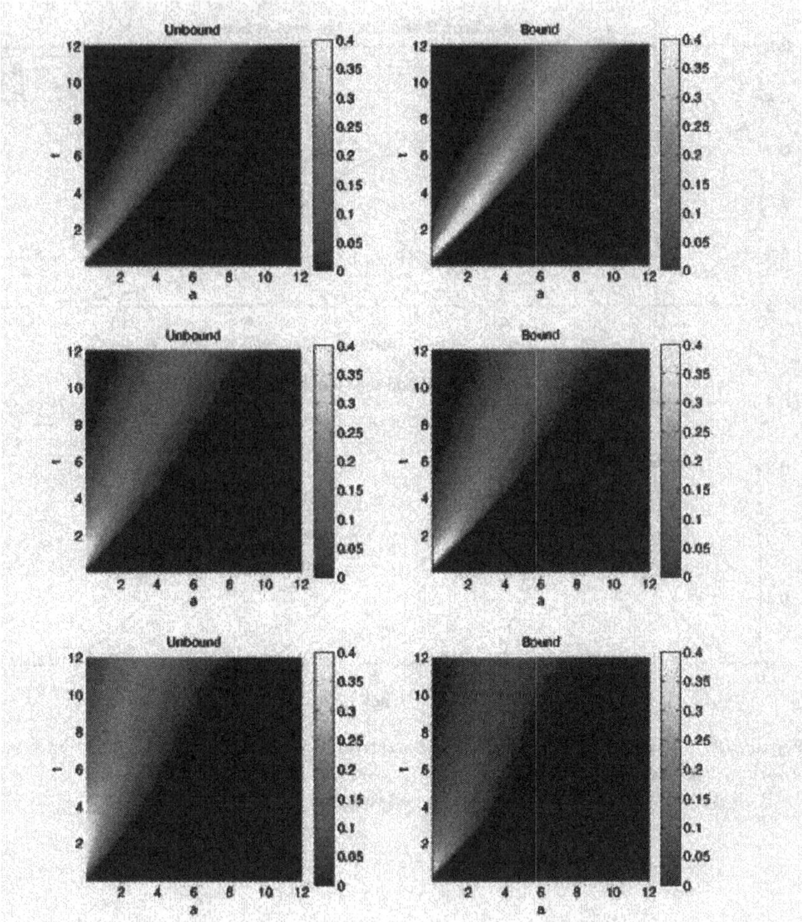

Figure 5. Solutions for different binding rates. The top row shows the solutions for U and B when $\alpha = 2$, middle row: $\alpha = 1$, and bottom row: $\alpha = .5$. In all figures $\lambda = 1$. The intensity of the color represents the amplitude of the solution. The peak amplitude moves along the a-axis at the rate $\frac{\alpha}{\alpha+\lambda}$.

of time cells spend in the bound state and cells only age when they are bound. So, the peak of the hump should move along (age) at a speed of $\frac{\alpha}{\alpha+\lambda}$. Furthermore, the peak of U lags behind the peak of B with a lag of $\frac{1}{\alpha+\lambda}$. This is because $\frac{1}{\alpha+\lambda} = \frac{1}{\alpha}\frac{\alpha}{\alpha+\lambda}$ which is the product of the mean time not bound and the time aging while bound. This is shown in Figure 5. Hence, we have characterized the peak speed and lag in this system, but have not yet solved for the variance in the system.

2.6. *Extending the solution: Allowing population growth*

The result in the simplified case is useful, but we would also like to solve the problem when γ is unequal to α and cells can divide.

Suppose the rate $p - k = \epsilon$ is small compared to α. Our system of equations to solve in characteristic coordinates is then

$$\frac{\partial U}{\partial \tau} = -(\alpha - \epsilon)U + \lambda B \tag{36}$$

$$\frac{\partial B}{\partial \eta} = \alpha U - \lambda B. \tag{37}$$

with

$$U(\eta, 0) = 0 \tag{38}$$

$$B(0, \tau) = \alpha u^* \exp(-\alpha \tau). \tag{39}$$

The boundary conditions remain the same and can be derived in a similar manner as described for the simpler problem. Adding the ϵ growth term affects only the internal dynamics of the system and does not alter what happens at the boundary. In biological terms, if cells can only divide once they have positive age, the cells of age zero will not be effected. The other boundary represents cells that have been bound the entire time of simulation. Bound cells cannot divide so again the addition of ϵ does not affect the boundary condition.

We use the same technique to solve, and find the Laplace transform in τ to obtain

$$s\widehat{U} - U(\eta, 0) = \lambda \widehat{B} - (\alpha - \epsilon)\widehat{U}$$

or

$$\widehat{U} = \frac{\lambda \widehat{B}}{(s + \alpha - \epsilon)}.$$

The equation for $\widehat{B}$ is

$$\frac{\partial \widehat{B}}{\partial \eta} = \alpha \widehat{U} - \lambda \widehat{B}.$$

with

$$\widehat{B}(0, s) = \frac{\alpha u^*}{\alpha + s}$$

exactly as before.

That is

$$\frac{\partial \widehat{B}}{\partial \eta} = \left[\frac{\alpha\lambda}{(s+\alpha-\epsilon)} - \lambda\right]\widehat{B} \Rightarrow \widehat{B} = \exp\left[\left(\frac{\alpha\lambda}{(s+\alpha-\epsilon)} - \lambda\right)\eta\right]\frac{\alpha u^*}{\alpha+s}$$

So, we must invert the transform:

$$\widehat{B} = \alpha u^* e^{-\lambda\eta}\frac{\exp\left[\frac{\alpha\lambda\eta}{s+\alpha-\epsilon}\right]}{\alpha+s}$$

but since the denominator of the argument of the exponential function does not match the denominator of the expression we cannot directly apply the rule we used before.

Provided that $\epsilon < \alpha/2$, we can write $\frac{1}{s+\alpha}$ as the sum of the infinite geometric series

$$\sum_{j=0}^{\infty} \frac{(-\epsilon)^j}{(s+\alpha-\epsilon)^{(j+1)}}.$$

Then we can write

$$\widehat{B} = \alpha u^* e^{-\lambda\eta}\sum_{j=0}^{\infty}\frac{(-\epsilon)^j \exp\left[\frac{\alpha\lambda\eta}{s+\alpha-\epsilon}\right]}{(s+\alpha-\epsilon)^{(j+1)}}$$

and

$$\widehat{U} = \lambda\alpha u^* e^{-\lambda\eta}\sum_{j=0}^{\infty}\frac{(-\epsilon)^j \exp\left[\frac{\alpha\lambda\eta}{s+\alpha-\epsilon}\right]}{(s+\alpha-\epsilon)^{(j+2)}}.$$

To invert this now use the same rule that we applied previously, namely:

$$f(s) = \frac{e^{-a/s}}{s^{n+1}}, \qquad n > -1$$

is the Laplace transform of

$$F(t) = \left(\frac{t}{a}\right)^{n/2} J_n(2\sqrt{at}).$$

So,

$$B(\eta,\tau) = \alpha u^* e^{-\lambda\eta} e^{-(\alpha-\epsilon)\tau}\sum_{j=0}^{\infty}(-\epsilon)^j\left(\frac{\tau}{-\alpha\lambda\eta}\right)^{j/2} J_j(2\sqrt{-\alpha\lambda\eta\tau})$$

or

$$B(\eta,\tau) = \alpha u^* e^{-\lambda\eta} e^{-(\alpha-\epsilon)\tau}\sum_{j=0}^{\infty}(-\epsilon)^j\left(\frac{\tau}{\alpha\lambda\eta}\right)^{j/2} I_j(2\sqrt{\alpha\lambda\eta\tau})$$

and

$$U(\eta,\tau) = \lambda\alpha u^* e^{-\lambda\eta} e^{-(\alpha-\epsilon)\tau} \sum_{j=0}^{\infty} (-\epsilon)^j \left(\frac{\tau}{-\alpha\lambda\eta}\right)^{(j+1)/2} J_j(2\sqrt{-\alpha\lambda\eta\tau})$$

or

$$U(\eta,\tau) = \lambda\alpha u^* e^{-\lambda\eta} e^{-(\alpha-\epsilon)\tau} \sum_{j=0}^{\infty} (-\epsilon)^j \left(\frac{\tau}{\alpha\lambda\eta}\right)^{(j+1)/2} I_j(2\sqrt{\alpha\lambda\eta\tau})$$

Let $x(\eta,\tau) = 2\sqrt{\alpha\lambda\eta\tau}$ and $W(\eta,\tau) = \alpha u^* e^{-\lambda\eta} e^{-(\alpha-\epsilon)\tau}$ and then

$$B(\eta,\tau) = W \sum_{j=0}^{\infty} (-\epsilon)^j \left(\frac{x}{2\alpha\lambda\eta}\right)^j I_j(x)$$

$$U(\eta,\tau) = \lambda W \sum_{j=0}^{\infty} (-\epsilon)^j \left(\frac{x}{2\alpha\lambda\eta}\right)^{j+1} I_{(j+1)}(x).$$

Back in original variables the exact solution for $0 < a < t$ is:

$$B(a,t) = \alpha u^* e^{-\lambda a} e^{-(\alpha-\epsilon)(t-a)} \times \sum_{j=0}^{\infty} (-\epsilon)^j \left(\frac{2\sqrt{\alpha\lambda a(t-a)}}{2\alpha\lambda a}\right)^j I_j(2\sqrt{\alpha\lambda a(t-a)}) \tag{40}$$

$$U(a,t) = \lambda\alpha u^* e^{-\lambda a} e^{-(\alpha-\epsilon)(t-a)} \times \sum_{j=0}^{\infty} (-\epsilon)^j \left(\frac{2\sqrt{\alpha\lambda a(t-a)}}{2\alpha\lambda a}\right)^{j+1} I_{(j+1)}(2\sqrt{\alpha\lambda a(t-a)}) \tag{41}$$

(Note that in the limit as $\epsilon \longrightarrow 0$, we have the same solution we had before.)

We can use the recurrence relation [4]

$$xI_{n+1}(x) = xI_n'(x) - nI_n(x)$$

to express the terms of this sum in terms of I_0 and derivatives of I_0. Therefore, we can see that if $\alpha\lambda\eta = d$,

$$B(\eta,\tau) = W \sum_{j=0}^{\infty} \left(\frac{-\epsilon x}{2d}\right)^j I_j(x) = W\left\{I_0(x) - \frac{\epsilon x}{2d} I_1(x) + O(\epsilon^2)\right\}$$

$$= W\left\{I_0(x) - \frac{\epsilon x}{2d} I_0'(x) + O(\epsilon^2)\right\} = WI_0(x+\epsilon y) + O(\epsilon^2)$$

with $y = \frac{x}{2d}$. Or in original coordinates for $0 < a < t$:

$$B(a,t) = \alpha u^* e^{-\lambda a} e^{-(\alpha-\epsilon)(t-a)} \times I_0\left(2\sqrt{\alpha\lambda a(t-a)} - \epsilon\sqrt{\frac{(t-a)}{\alpha\lambda a}}\right) + O(\epsilon^2). \tag{42}$$

Using this recurrence we can also write U in simplified form as follows:

$$\begin{aligned}
U(\eta,\tau) &= \lambda W \sum_{j=0}^{\infty} (-\epsilon)^j \left(\frac{x}{2d}\right)^{j+1} I_{j+1}(x) \\
&= \lambda W \left\{ \frac{x}{2d} I_1(x) - \frac{\epsilon x^2}{4d^2} I_2(x) + O(\epsilon^2) \right\} \\
&= \lambda W \left\{ \frac{x}{2d} I_1(x) - \frac{\epsilon x}{4d^2} (x I_1'(x) - I_1(x)) + O(\epsilon^2) \right\} \\
&= \frac{\lambda W}{2d} \left\{ \left(x + \frac{\epsilon x}{2d} \right) I_1(x) - \frac{\epsilon x^2}{2d} I_1'(x) + O(\epsilon^2) \right\} \\
&= \frac{\lambda W}{2d} \left\{ (x + \epsilon y)(I_1(x - \epsilon y)) \right\} + O(\epsilon^2)
\end{aligned}$$

Thus, in original coordinates for $0 < a < t$:

$$\begin{aligned}
U(a,t) = \frac{u^*}{2a} e^{-\lambda a} e^{-(\alpha-\epsilon)(t-a)} &\left(2\sqrt{\alpha\lambda a(t-a)} + \epsilon\sqrt{\frac{(t-a)}{\alpha\lambda a}} \right) \\
\times I_1 &\left(2\sqrt{\alpha\lambda a(t-a)} - \epsilon\sqrt{\frac{(t-a)}{\alpha\lambda a}} \right) + O(\epsilon^2). \qquad (43)
\end{aligned}$$

3. Simulation Results

The discussion of this section refers to solutions of the full model of Section 2. Using our numerical scheme above, we investigate the model predictions for some biological questions:

- How is the resulting age distribution of a single clonal line affected by changes in the parameters?
- What is the effect when antigen is scarce?
- If memory cells are intermediates, how can memory be sustained?
- Do higher affinity clones produce more activated cells?

3.1. *Parameter selection and assumptions*

We solve equations 1–6 with the nonconstant binding rate α and proliferation rate p. This full system contains only six parameters. The precise assumptions we make for our numerical simulations are as follows:

- We assume that proliferation begins only after cells have reached the age of $a_p = .83$ day ≈ 20 hours. (All cells of that age and beyond proliferate with rate $\widehat{p}$, regardless of antigen affinity.)

- Proliferation of unbound cells happens at a rate of $\widehat{p} = 2.4$ per day, equivalent to every 10 hours on average.
- Cells leave the lymph node at a constant rate of $k = .2$ per day, or roughly every 5 days on average.
- Total antigen sites available on dendritic cells is $S = 100$. This value is somewhat arbitrary since the model can essentially be scaled by the available antigen sites, but we choose 100 for convenience.
- The chance of encountering a binding site if all possible sites are available is $\widehat{\alpha} = 48$ per day which means, roughly every half-hour. (We also run simulations with other values of this parameter).
- The expected time to unbind for a particular T clone depends on the affinity of that clone for the antigen. We look at $\lambda = 72$, 48, and 24 per day, corresponding to 20 minutes, 30 minutes, and 1 hour respectively for the average time to dissociation. A low value of λ corresponds to a high affinity clone.

3.2. *Single clone results*

A numerical solution to a single clone simulation with one initial naive cell as the initial condition is displayed in Figure 7. The top two plots show contour lines for $U(a,t)$ and $B(a,t)$. Very rapidly the simulation reveals smooth peaks that age at a rate $\frac{\alpha}{\alpha+\lambda}$. The amplitude first decays, then grows, and the peaks spread as they evolve. The bottom plot of this figure illustrates the age distribution of activated cells that have entered into circulation. After 4 days a substantial number of activated cells are released — perhaps these cells can be interpreted as M_0 or A_0 of the previous chapter. These cells, which have surpassed an activation threshold (some fixed age), are primed to respond at a site of inflammation in the tissues.

We numerically solved the problem for a range of values of $\widehat{\alpha}$ and λ, and show the resulting age distributions in Figure 6. Observe that the spread of the resulting age distribution is wider when $\widehat{\alpha}$ and λ are smaller. An interesting observation is that the high affinity clones do not supply the greatest number of activated clones. In fact, since cells cannot divide while they are bound, a clone with very high affinity will age too quickly before having the opportunity to proliferate much.

On the other hand, if the binding rate is low, then low affinity clones do not age quickly enough to reach a proliferating age before the end of the simulation. So, assuming that the end of simulation corresponds with when antigen is no longer displayed, these clones will not produce many

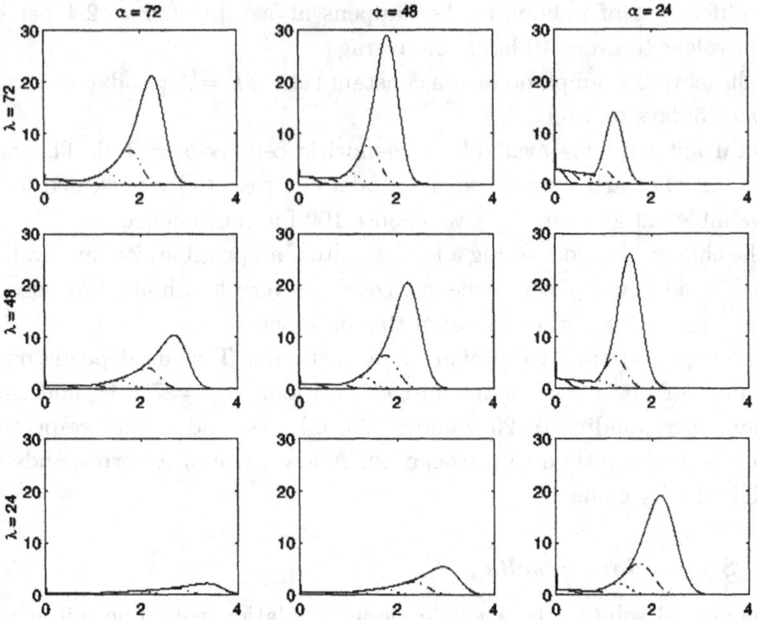

Figure 6. Cells in circulation for ranges of binding/dissociation rates. The plots show the age distribution (predicted by the model) of cells that have left the lymph node after 1, 2, 3, 4, and 5 days of activation by mature dendritic cells. The amplitude is different depending on the values of $\widehat{\alpha}$ and λ. High values of $\widehat{\alpha}$ indicate that T cells make frequent encounters with the antigen sites. Low values of λ are associated with high affinity of the clone to the antigen:MHC.

activated cells either. Therefore, we have identified a trade–off: low affinity clones will not age fast enough to reach the age of proliferation, but the highest affinity clones will stay bound and thus be less likely to proliferate. Indeed, the peaks shown in Figure 6 suggest that there is an optimum ratio of $\widehat{\alpha}$ to λ that will produce the most cells. It appears that the value of $\widehat{\alpha}/\lambda$ is less than one. This could be something to verify experimentally, or it could be a flaw in the model.

3.3. *Scarcity of antigen*

When antigen presentation sites are not numerous, encountering an available site becomes difficult early on in the simulation. Sites become rapidly occupied, and so the binding rate α in equation 3 approaches zero. This means that the speed of aging ($\frac{\alpha}{\alpha+\lambda}$) also rapidly diminishes. Unbound

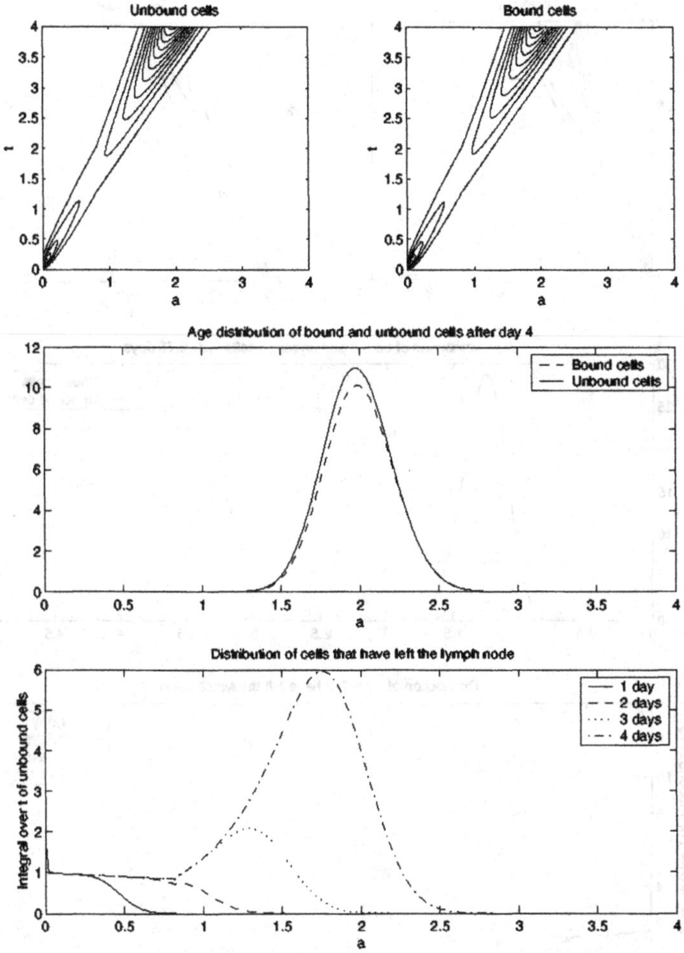

Figure 7. Single clone simulation. Numerical results for the parameter values $\lambda = 48$ per day, $\widehat{\alpha} = 48$ per day, $k = .2$ per day, $\widehat{p} = 2.4$ per day and $S = 100$. The top row shows the contours of U and B with age along the horizontal axis and time increasing up the vertical axis. The center plot shows the cross section of those results at $t = 4$, and the bottom plot shows a measure of the age distribution of cells that have left the lymph node after 1 day, 2 days, 3 days and 4 days.

cells that have surpassed age a_p continue to proliferate, but aging is stalled because α is near zero. Hence, the model suggests more cells proliferate at a lower age when antigen is scarce. Characteristic solutions for $S = 1$ that demonstrate this effect are shown in Figure 8.

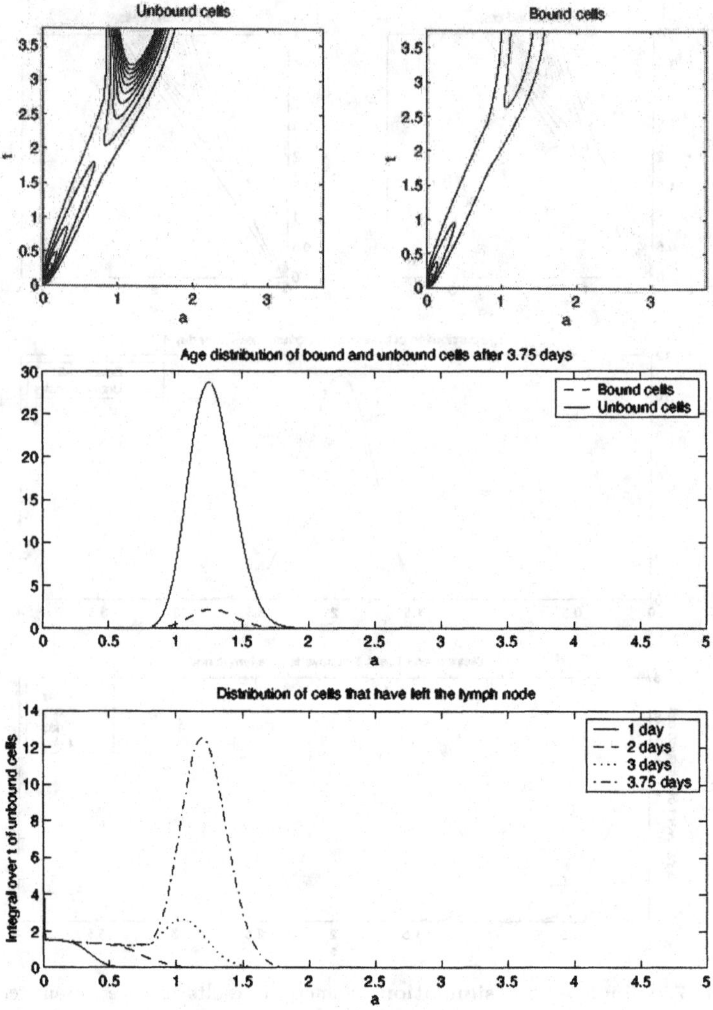

Figure 8. Results with limited antigen. Numerical results as in Figure 7 for the parameter values $\lambda = 48$ per day, $\widehat{\alpha} = 48$ per day, $k = .2$ per day, $\widehat{p} = 2.4$ per day and $S = 1$. The scarcity of antigen causes the aging to stall.

3.4. *Memory*

Perhaps the most obvious criticism of a model that assumes memory cells are intermediates is, one could argue, that such a model would not be capable of sustaining lifelong memory. In this model, memory cells can only age their way towards activation induced cell death, and cannot move backwards in the differentiation process. How, then, can memory be replenished when antigen is encountered serially?

Indeed, if cells remained indefinitely in the lymph node, the model predicts that eventually intermediate cells would disappear. The support of the solution must move to the right along the age axis. The previous section points out, however, that aging can become slower or stall when antigen sites are occupied, allowing activated cells to proliferate and generate more copies of cells with the same level of maturity. This suggests one mechanism for maintaining intermediates in cases of chronic infection.

However, more importantly, cells leave the lymph node. The necessity of local interactions means that removed cells intrinsically store and retain memory without aging to death. We assume that IL-2 is not available for reuptake while cells circulate in the blood. Until they re-encounter antigen, perhaps at the site of infection, proliferation is thereby halted.

We use the following simulation to illustrate that memory is maintained and a multitude of effectors are generated quickly with subsequent infections. We assume at the first infection the lymph node displays antigen for 7 days. The resulting distribution of cells is given by some multiple of the integral of U and we assume this is enough to wipe out the infection. To simulate a secondary response, we then take a fraction (0.10, assuming 10% of the cells survive long term) of that resultant distribution as our initial condition and see what happens after 3 days of antigen presentation in the lymph node. This simulates a response to a secondary infection. Our simulation result is shown in Figure 9. Supposing that the memory cells are those with ages between 1 and 3, while effectors are cells with ages between 3 and 5, the contour plots show what is expected of secondary immune responses. Numerous effectors are generated quickly (presumably enough to clear the infection) and intermediate cells persist. This analysis assumes nothing significant happens outside the lymph node, other than an overall population reduction. However, more precise models of the effects of antigen encounter at an infection site will be the topic of future research.

This type of simulation could be repeated where resulting distributions are used as initial data. Over time is there a persistent memory or do the intermediate cells eventually die off? The answer depends on the parameters of simulation. If they eventually die off then we have a possible explanation for antigenic shift in chronic or persistent infections or autoimmunity. If not, then we have an mechanistic explanation for stable memory maintenance. Figure 10 shows the net cell distributions after a first, second and third encounter with antigen. The assumptions of this simulation are that again the first infection lasts for 7 days while the subsequent infections last 3 days.

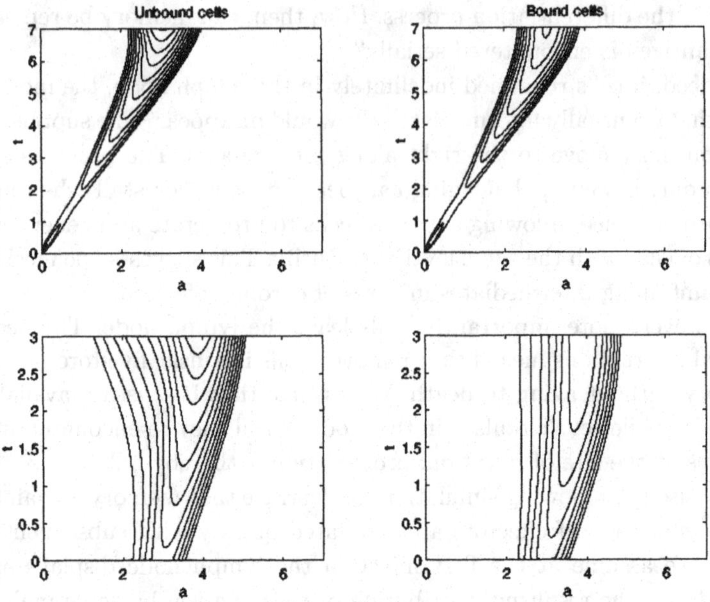

Figure 9. Simulation of primary and secondary infections. This figure shows the solutions for T population distributions after primary and secondary challenges. The top row depicts contour plots after a primary infection, and the second row shows the contours after a secondary challenge.

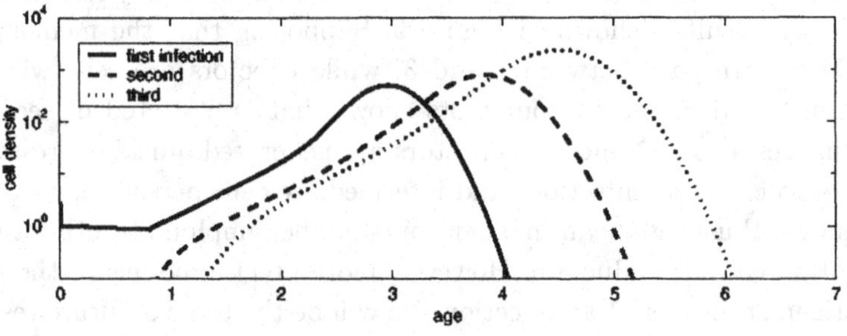

Figure 10. Simulation of repeated infections. This figure shows the resulting age distributions of cells that emerge from the lymph nodes after a first, second and third infection. We assume that antigen is presented for 7 days in the primary infection, and for only 3 days in the subsequent challenges.

3.5. *Two clones competing for antigen sites*

Another interesting topic for investigation are the competitive effects between clones of T cells with different antigen affinities. Our model predicts that in some simulations, the highest affinity clone is perhaps not the fittest due to the observations mentioned in Section 3.2. Typically the higher affinity clone initially out-competes the low affinity clone because its population ages faster and so high affinity clones reach maturity faster and can begin proliferation. Interestingly, however, low affinity clones can catch up to the high affinity ones and once they have reached maturity they can actually proliferate more since they spend less time bound to antigen on dendritic cells. We illustrate this phenomenon in Figure 11.

Are the highest affinity clones always selected? To explain this, we might need additional mechanisms — for example, up to now we have left out any age dependence of the binding probability α or k. It could be that as cells reach maturity they are less inclined to bind to the antigen presented in the

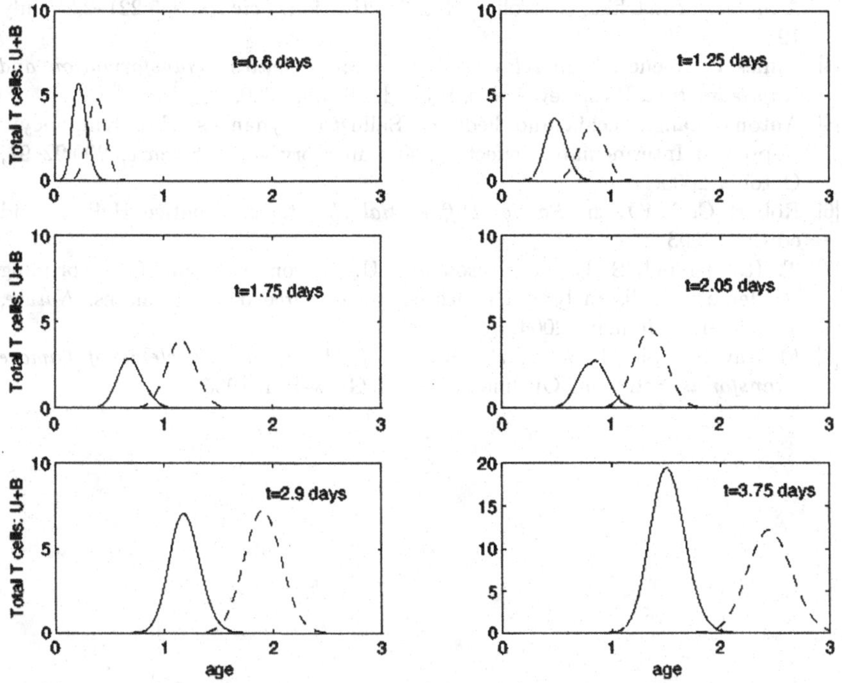

Figure 11. Two clone simulation. The frames display the population distributions of two clones of T cells at various times after entering the lymph node together. The high affinity clone (shown with a dotted line) is simulated with $\lambda = 24$ and the low affinity clone (solid line) has $\lambda = 72$. Although in the first couple of days, the high affinity clone matures faster and proliferates more, after the third day, the low affinity clone's population dominates the response. Notice the scale change in the vertical axis of the last frame. Other parameters are assigned the same values as in Figure 7.

lymph node. They are known to express surface proteins which make them likely to leave the lymph node and home to tissues. In this case the highest affinity clone could possibly restore its advantage. These precise workings are quite interesting topics of further research.

References

[1] Philippe Bousso and Ellen Robey. Dynamics of CD8+ T cell priming by dendritic cells in intact lymph nodes. *Nature Immunology*, 4(6):579–585, June 2003.

[2] Byron Goldstein, James R. Faeder, and William S. Hlavacek. Mathematical and computational models of immune–receptor signaling. *Nature Reviews Immunology*, 4:445–456, June 2004.

[3] Arash Grakoui, Shannon K. Bromley, Cenk Sumen, Mark M. Davis, Andrey S. Shaw, Paul M. Allen, and Michael L. Dustin. The immunological synapse: A molecular machine controlling T cell activation. *Science*, 285:221–227, July 1999.

[4] James P. Keener. *Principles of Applied Mathematics: Transformation and Approximation.* Westview Press, revised edition, 2000.

[5] Antonio Lanzavecchia and Federica Sallusto. Dynamics of T lymphocyte responses: Intermediates, effectors, and memory cells. *Science*, 290:92–97, October 2000.

[6] Robert C. McOwen. *Partial Differential Equations.* Prentice Hall, second edition, 2003.

[7] T. R. Mempel, S. E. Henrickson and U. H. von Andrian. T-cell priming by dendritic cells in lymph nodes occurs in three distinct phases. *Nature*, 427:154–159, January 2004.

[8] Murray R. Spiegel. *Schaum's Outline of Theory and Problems of Laplace Transforms.* Schaum's Outline Series. McGraw-Hill, 1965.

MODELING AND SIMULATION OF AGE- AND SPACE-STRUCTURED BIOLOGICAL SYSTEMS

BRUCE P. AYATI

Department of Mathematics
Southern Methodist University
Dallas, TX 75205
ayati@smu.edu

This expository paper provides an introduction to computational methods for solving continuous models of certain biological systems. These systems have important behavior in space and within the age distribution of individuals. Very often, the spatial behavior depends critically on the age structure of the population. The different spatial scales induce different time scales in the problem, whereas age and time advance together.

Age- and space-structured multiscale systems arise in a wide variety of biological problems ranging from multicellular and tissue-level phenomena to problems in ecology and evolutionary biology. We describe a general modeling framework used to represent such biological systems. We then discuss computational methods used to solve the model equations, beginning with a treatment of the moving-grid Galerkin method used to decouple age and time while they advance together, introduce how this discretization in age works with discretizations in time and space, and then review how more complicated nonlinear problems would be treated. We close by presenting two example systems, swarm-colony development of the bacteria *Proteus mirabilis* and tumor invasion, which differ in some important respects from the general model, but which are effectively treated by the computational methods presented in this paper.

Keywords: computation, age structure, *Proteus mirabilis*, tumor invasion, cell cycle.

1. Introduction

In this expository paper we provide an introduction to the modeling and simulation of certain biological systems where important behavior occurs in space and in the age demographics of the individual organisms or cells within the system. Such systems arise in the study of multicellular and tissue-level phenomena, ecology, epidemiology, and population genetics. Excellent surveys of physiologically structured models, including age-structure, can be found in [9, 20, 36].

The general class of partial differential equations for diffusion and age structure considered in this paper has a long history. Among the first classic works are Skellam (1951) [33] (who considered the effects of diffusion on populations), and Sharpe and Lotka (1911) [32] and McKendrick (1926) (who considered population models with linear age structure) [26, 36]. Later, Gurtin and MacCamy [17] considered models with nonlinear age structure. Rotenberg [31] and Gurtin [16] posed models dependent on both age and space. Gurtin and MacCamy [18] differentiated between two kinds of diffusion in these models: diffusion due to random dispersal, and diffusion toward an area of less crowding. Existence and uniqueness results can be found for various forms of these models in Busenberg and Iannelli [8], di Blasio [13], di Blasio and Lamberti [14], Langlais [22], MacCamy [24], and Webb [37]. Further analysis has been done by several authors [19, 21, 23, 25].

The organization of the paper is as follows. We discuss a general partial different equation that possesses the most common and most important features used to model age- and space-structured biological systems, followed by the computational method for solving the general equation in age, including how one would view the the method in the context of a Leslie matrix model. We then present one of the more straightforward implementations of a fully discrete method in age, time, and space and summarize the convergence results published in the numerical analysis literature. We close by reviewing two biological systems where both age and space structure are important. In the first system, *Proteus mirabilis* swarm-colony development, the spatial-temporal dynamics are in fact an expression of their unique lifecycle. In the second system, tumor invasion, age-structure is important for a mechanistic representation of cell growth and division within a spatially explicit model.

2. General Problem

In this section we derive a general model for the dynamics of a population distributed in age and space. We begin by ignoring any behavior in space and assume a constant death rate $\mu > 0$. The relationship between the age distribution at time t and time $t + \Delta t$ is given by the difference equation

$$u(a + \Delta t, t + \Delta t) = u(a, t) - \mu u(a, t)\Delta t. \tag{2.1}$$

We can write this relationship as

$$\frac{u(a+\Delta t, t+\Delta t) - u(a,t)}{\Delta t} = -\mu u(a,t). \tag{2.2}$$

Adding and subtracting $u(a, t+\Delta t)$ in the numerator of the left-hand side gives

$$\frac{u(a+\Delta t, t+\Delta t) - u(a, t+\Delta t)}{\Delta t} + \frac{u(a, t+\Delta t) - u(a,t)}{\Delta t} = -\mu u. \tag{2.3}$$

As $\Delta t \to 0$, we get $u_t + u_a = -\mu u$. In this paper, a subscript which is an independent variable in the problem denotes partial differentiation by that variable. Taking $\Delta t \to 0$ in Equation (2.2) gives $\mathcal{D}u = -\mu u$ where $\mathcal{D}$ is a total derivative in age and time. Although age-structured partial differential equations are by convention written using the sum of the age and time partial derivatives, the total derivative better suggests how we should approach solving the equation.

If we explicitly include the ability of individuals to move in space, our model equations may take the form

$$u_t + u_a = \mathcal{L}(u), \tag{2.4}$$

where $\mathcal{L}$ is an elliptic operator and $u(x, a, t)$ is a population distributed in space x, age a, and time t. In models with only one spatial dimension, we often have $\mathcal{L}(u) = ku_{xx} - \mu u$. For systems of equations, the elliptic operator may include additional zero-order reaction terms. To be well posed, Equation (2.4) requires suitable initial and boundary conditions. In particular, we require a condition at $u(x, 0, t)$, called the birth condition. If individuals are born at age 0 from parents older than themselves, we get a nonlocal boundary condition such as

$$u(x, 0, t) = \int_0^\infty \beta(a) u(x, a, t)\, da, \tag{2.5}$$

where $\beta(a)$ is the birth rate, or fecundity, at age a. If u is a population of cells where two daughter cells are created through mitosis from one mother cell, our nonlocal boundary condition can be refined to the form

$$u(x, 0, t) = 2 \int_0^\infty \theta(a) u(x, a, t)\, da, \tag{2.6}$$

where $\theta(a)$ is the cell-division rate and the splitting constant 2 has been made explicit. Depending on what age represents in a given model, we may have other forms of the "birth" condition, such as Equation (4.32d) in the model of *Proteus mirabilis* presented in Section 4.1.

A common general nonlinear model for age- and space-structured systems, and one used for the original development and analysis of the numerical methods discussed in this paper [3, 5], is as follows. The population $u(x,a,t)$ is distributed as above, although space may have more than one dimension. The age- and space-structured partial differential equation is

$$u_t + u_a = \nabla \cdot k(x,p)\nabla u - \mu(x,a,p)u, \quad x \in \Omega,\ a > 0,\ t > 0, \tag{2.7}$$

where $\Omega \subset \mathbb{R}^n$ represents the spatial domain. The diffusion, $\nabla \cdot k\nabla u$, arises from the symmetric random motion of each individual (Fickian diffusion). Here ∇ and $\nabla\cdot$ denote the gradient and the divergence, respectively, in x. Isotropic random motion results in diffusion of the form $\nabla^2(ku)$. The choice between diffusions should be based on biological considerations. See [2, 28, 29, 35] for discussions and derivations of different diffusions and other continuous representations of taxis.

The total population density, p, is given by

$$p(x,t) = \int_0^\infty u(x,a,t)\,da, \quad x \in \Omega,\ t > 0. \tag{2.8}$$

We have a birth condition

$$u(x,0,t) = \mathcal{B}(x, u(x,\cdot,t)), \quad x \in \Omega,\ t > 0, \tag{2.9}$$

that is dependent on the entire population distribution. We note that $\mathcal{B}$ is an operator whose second argument is a function defined on $\mathbb{R}^+$, where $\mathbb{R}^+$ denotes the non-negative real numbers. We have a Neumann boundary condition, with ν denoting the outward normal to the boundary $\partial\Omega$,

$$k(x,p)\nabla u \cdot \nu = 0, \quad x \in \partial\Omega,\ a > 0,\ t > 0, \tag{2.10}$$

that represents an isolated habitat. The initial condition is

$$u(x,a,0) = u_0(x,a), \quad x \in \Omega,\ a > 0. \tag{2.11}$$

The model in equations (2.7)–(2.11) contains many, if not most, of the features and complications that are expected to rise in the application of this modeling framework to real biological problems. Numerical methods designed to to solve this general model efficiently and robustly can be expected to be effective in solving more biologically realistic models that deviate in some of the mechanisms. This has proven to be the case in two example systems discussed below, swarm-colony development of the bacteria *Proteus mirabilis* (Section 4.1) and tumor invasion (Section 4.2).

3. Computational Methods

In this section we provide a treatment of what amounts to a fairly complex computational framework, but is hopefully more accessible to mathematical biologists than what has been published in the numerical analysis literature. We begin with a discussion of the moving-grid Galerkin method that forms the core of the computational methodology. However, in the interests of maximizing accessibility, we discuss a formulation using piecewise-constant functions as the approximation space in age, originally developed in [3]. This is the lowest order approximation space possible, but has proven to be effective in practice. It has the additional benefit of lending itself to an explanation that does not require explicit reference to a Galerkin formulation. This is not the case of the higher-order approximation spaces in age developed and treated in [5]. We relate the moving-grid Galerkin method to Leslie matrix models in the hope of increasing intuition in the former. We then discuss one way the moving-grid Galerkin method can be combined with methods in time and space to solve systems similar to Equations (2.7)–(2.11). We end the section with a discussion of the convergence results and their implications.

Some readers may be familiar with the Escalator Boxcar Train (EBT), a method with a probabilistic derivation developed by de Roos [10, 11] for more general physiologically-structured population models that do not involve spatial structure. The method can also be viewed as a generalized Leslie matrix model and involves moving the discretization along characteristic curves. The formulation of de Roos' method leaves time continuous and thus any suitable discretization of time will be independent from the age or size discretization. There are differences from the methods presented in this paper in the handling of birth, death, the representation of the approximate solution, and generalizations to higher order methods. We are also unaware of any completed convergence analysis of the EBT method as sought by de Roos and Metz [12].

3.1. *Moving-grid Galerkin method in age*

Our approach to solving Equations (2.7)–(2.11) in age is essentially one of discretizing the total derivative $\mathcal{D}$. Solving a first-order hyperbolic term along a total derivative in this manner is referred to as a method of characteristics. The characteristic curves in this case are the lines in the age-time plane with slope equal to one, reflecting the fact that age and time advance together. A numerical method that solves the problem along the age and time characteristics will be free of numerical dispersion, a potential source of significant numerical error. This is not generally the case with other discretizations (see Chapter 7 of [34] for a discussion).

The discretization on the left-hand side of Equation (2.2) is the simplest means of solving a system along age-time characteristic lines, but contains a crippling constraint: the age and time steps are constant and must equal one another. This constraint is a consequence of any fixed discretization, illustrated in Figure 1(a). In models where there are also spatial dynamics, the time step needed to solve the system accurately in space will be much smaller than what is needed for an accurate age discretization, typically inflating the computational cost by a factor of 10 to 100. Given the already high computational cost of solving systems with a large number of dimensions, this added and unnecessary expense can render problems unsolvable in practice. Moreover, the requirement that the time steps remain fixed in length prevents adaptivity in time, an important means of gaining computational efficiency without loss of accuracy.

The solution to the dilemma of solving along characteristics so as to advance age and time together, while decoupling the age and time discretizations, is to use an age discretization where the discrete age intervals move along characteristic lines, illustrated in Figure 1(b). The moving grid Galerkin methods described in this section contains the method of characteristics on a fixed grid as a special case.

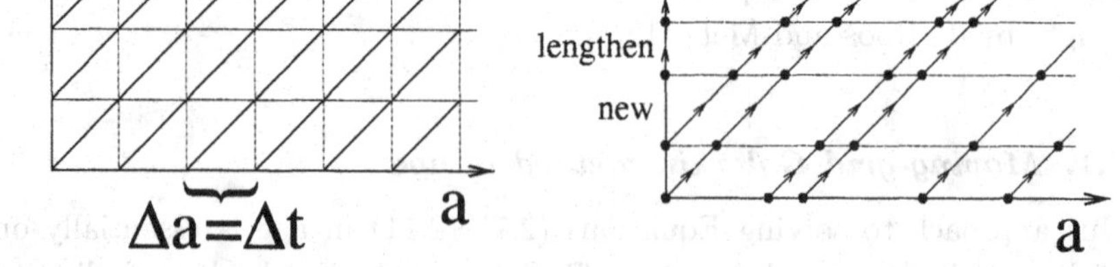

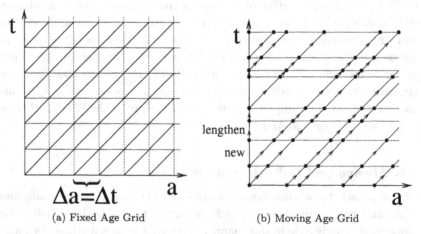

(a) Fixed Age Grid (b) Moving Age Grid

Figure 1. (a) Use of a fixed age grid with a method of characteristics imposes the requirement that the age steps and time steps are equal and constant. (b) Use of a moving age grid allows movement along characteristics even with different and nonuniform age and time steps.

The moving grid accounts for the aging of the population as time progresses. The main complication from using a moving grid is in the nature of the first age interval. If we introduce a new age discretization point (node) at age 0 every time step, then we have recoupled the age and time discretizations (although we now have the capacity for variable time steps). Instead we must allow for the first age interval (birth interval) to lengthen during certain time steps.

To motivate our handling of the lengthening of the first age interval, consider a situation where we ignore death and diffusion, and assume a influx of newborns, b. Let U denote the population density in the first age interval. The subscripts "new" and "old" denote the end and beginning of the time step, respectively. Over a time step Δt, we have a conservation law

$$\Delta a_{\text{new}} U_{\text{new}} = \Delta a_{\text{old}} U_{\text{old}} + \Delta t\, b. \tag{3.12}$$

Note that $\Delta a_{\text{new}} = \Delta a_{\text{old}} + \Delta t$ so that $\Delta a_{\text{old}}(U_{\text{new}} - U_{\text{old}}) + \Delta t U_{\text{new}} = \Delta t\, b$. Then

$$\frac{U_{\text{new}} - U_{\text{old}}}{\Delta t} = \frac{b - U_{\text{new}}}{\Delta a_{\text{old}}}. \tag{3.13}$$

The left-hand side is a difference approximation to the time derivative and b in the numerator of the right-hand side accounts for the new individuals born into the first age interval. The term $-U_{\text{new}}$ in the numerator of the right-hand side accounts for the decrease in population density due to conservation of population in the context of a lengthening age interval. Equation (3.15) encompasses a more rigorous treatment of the penalty term for the lengthening of the first age interval. A similar penalty term occurs in the higher-order versions of the methods, although it is couched in the notation of a Galerkin method [5].

The moving grid in age has an intuitive demographic meaning. Assume Betty was born in March 1971. In February 2006, she will be classified as 34 years old, corresponding to a fixed age discretization. However, assuming Betty grew up in the United States, it is likely that she graduated from high school in 1989 and can be referred to as belonging to the Class of '89. Sometime in March 2006, Betty will be reclassified as 35 years old and change cohorts. However, she will remain a member of the Class of '89 for her entire life, as will her classmates born a few months earlier or later. In contrast, with the standard notion of an age group, these classmates will belong to different cohorts as the year progress. The class notion of a cohort underlies the discretizations used in this paper, and this example retains a critical complication of the method. In February 2006, the label

"Class of 2024" defines an age group that is still in the process of being formed, whereas the category of children under the age of one remains as it has always been defined. However, once an interval is fully formed, no additional computation is needed to figure out who has moved from one fixed age group to the next.

More rigorously, the moving age grid constitutes a transformation of Equations (2.7)–(2.11) to a moving reference frame in age, $w(x,c,t) = u(x, c+t, t)$, resulting in the system

$$w_t = \nabla \cdot k(x,p)\nabla w - \mu(x, c+t, p)w, \quad x \in \Omega,\ c > -t,\ t > 0, \tag{3.14a}$$

$$w(x,-t,t) = \mathcal{B}(x, w(x,\cdot - t, t)), \quad x \in \Omega,\ t > 0, \tag{3.14b}$$

$$k(x,p)\nabla w \cdot \nu = 0, \quad x \in \partial\Omega,\ c > -t,\ t > 0, \tag{3.14c}$$

$$p(x,t) = \textstyle\int_{-t}^{\infty} w(x,c,t)\, dc, \quad x \in \Omega,\ t > 0, \tag{3.14d}$$

$$w(x,c,0) = u_0(x,c), \quad x \in \Omega,\ c > -t. \tag{3.14e}$$

In order to define well the introduction of new age nodes, we discretize age for all time by using an infinite mesh, $c_{-\infty} < \cdots < c_{-1} < c_0 < c_1 < \cdots c_{\infty}$. Each age interval $[c_i, c_{i+1})$, $-\infty < i < \infty$, can be thought of as defining a cohort. The infinite mesh is a useful analytic tool. The actual c_i's may be determined adaptively as the calculation progresses. We denote the partition determined by the set of points $\{c_i\}$ by γ. If $c_{i+1} > -t$, the *active cohort* is defined by the interval $\mathcal{C}_i(t) = [\max(c_i, -t), c_{i+1})$, with length $\Delta c_i(t)$. We denote by $\mathcal{J}(\gamma, t)$ the space of functions which are constant over each $\mathcal{C}_i$; extended to be zero for $c < -t$. We let W_i denote the age-discrete approximation to w on $\mathcal{C}_i$ and $W(x,t)$ denote the corresponding function in $\mathcal{J}(\gamma,t)$. In practice we need only consider W_i for c_{i+1} in the interval $[-t, a_{\max} - t]$, for a suitably large choice of $a_{\max}$. A discontinuous piecewise-polynomial over a fixed grid in a moving reference frame in age is illustrated in Figure 2.

Let A denote the age-averaging operator.[a] To motivate the birth and death terms in the definition of the age-discrete model, we apply the age-averaging operator to Equation (3.14a). We note that if $-t \in \mathcal{C}_i(t)$, then $\frac{\partial}{\partial t}\Delta c_i(t) = 1$. We can then apply the product rule to $\frac{\partial}{\partial t}\mathsf{A}_i(t,w)$ to obtain

$$\mathsf{A}_i(t, w_t) = \frac{\partial}{\partial t}\mathsf{A}_i(t,w) + \frac{1}{\Delta c_i}\big(\mathsf{A}_i(t,w) - \mathcal{B}(x,w)\big). \tag{3.15}$$

---

[a]Let χ denote the characteristic function. Formally, we define $\mathsf{A} : L^1_{\text{loc}} \to \mathcal{J}$ by $\mathsf{A}(t,\varphi)(c) = \sum_i \mathsf{A}_i(t,\varphi)\chi_{\mathcal{C}_i(t)}(c)$, where $\mathsf{A}_i(t,\varphi) = \frac{1}{\Delta c_i(t)}\int_{\mathcal{C}_i(t)} \varphi(c)\, dc$. The operators A and A_i will be applied only to the variable c in $\mu(x, c+t, p)$.

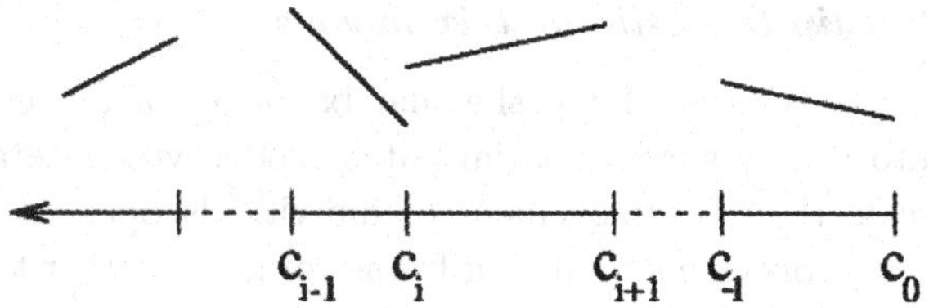

Figure 2. A discontinuous piecewise polynomial over a fixed grid in a moving reference frame. Shown are discontinuous piecewise-linear functions. Indexing begins on the right at zero and continues to negative infinity. At any given point in time, we only need to consider individuals whose age in the moving reference frame lies within $[-t, a_{\max} - t]$, for a suitably large choice of $a_{\max}$. This corresponds to the normal age domain of $[0, a_{\max}]$.

For the death term, we have

$$\begin{aligned} \mathsf{A}_i(t, \mu w) &= \mathsf{A}_i(t,\mu)\mathsf{A}_i(t,w) + \mathsf{A}_i(t, (\mu - \mathsf{A}_i(t,\mu))w) \\ &= \mathsf{A}_i(t,\mu)\mathsf{A}_i(t,w) + \mathsf{A}_i(t, (\mu - \mathsf{A}_i(t,\mu))(w - \mathsf{A}_i(t,w))). \end{aligned} \tag{3.16}$$

Thus $\mathsf{A}_i(t,\mu)\mathsf{A}_i(t,w)$ is a second-order correct approximation of $\mathsf{A}_i(t,\mu w)$, if μ and w are smooth, since $\mu - \mathsf{A}_i(t,\mu)$ and $w - \mathsf{A}_i(t,w)$ are both of first-order in age. We will obtain a second-order correct model in age by using a parabolic equation for the population density of each cohort,

$$\frac{\partial}{\partial t} W_i(x,t) = \nabla \cdot k(x,P)\nabla W_i(x,t) - \bar{\mu}_i(x,t,P)W_i(x,t) + B_i(x,t,W), \tag{3.17}$$

for every i such that $c_{i+1} > -t$. If $c_{i+1} = -t$, then $W_i = \mathcal{B}(x,W)$. The death modulus is

$$\bar{\mu}_i(x,t,\varphi) = \mathsf{A}_i\big(t, \mu(x, \cdot + t, \varphi)\big). \tag{3.18}$$

The birth term is

$$B_i(x,t,\varphi) = \begin{cases} \frac{1}{\Delta c_i}\left(\mathcal{B}(x,\varphi) - \mathsf{A}_i(t,\varphi)\right), & \text{if } -t \in \mathcal{C}_i(t), \\ 0, & \text{otherwise.} \end{cases} \tag{3.19}$$

We take $W_i = 0$ for $c_{i+1} < -t$. The term $\mathsf{A}_i(t, W_i)/\Delta c_i$ in $B_i(x,t,W)$ accounts for the conservation of population as the length of the active birth interval increases. Notice that if we temporarily neglect the k and μ terms, then

$$\frac{\partial}{\partial t}\left(\Delta c_i W_i\right) = \mathcal{B}(x, W_i), \quad -t \in \mathcal{C}_i(t). \tag{3.20}$$

The age-discrete total population density, P, is obtained by integrating W in the age variable. The equations are coupled only through B and P.

3.2. *Relationship to Leslie matrix models*

In this section we embed the Leslie matrix model of an age-structured population into a very simple spatial patch model with interactions on a faster time scale than the movement of individuals between age groups. The goal is not to provide a modeling framework, but rather to build some intuition for the idea of a moving age grid in the context of the continuous models covered in the other sections in this paper.

The Leslie matrix model (following Chapter 22 in [20]) consists of a discrete number of age groups or cohorts, represented by an m-vector

$$\mathbf{p}(t) = \begin{bmatrix} p_0(t) \\ p_1(t) \\ \vdots \\ p_{m-1}(t) \end{bmatrix}, \tag{3.21}$$

whose evolution is tracked in discrete time, t. Space is not considered. The time-step-to-time-step survival probabilities for each cohort are given by P_a for $a = 0, \ldots, m-2$. Fecundities are given by F_a for $a = 0, \ldots, m-1$. The Leslie matrix model is the equation $\mathbf{n}(t+1) = L\mathbf{n}(t)$ where the Leslie matrix is

$$L = \begin{bmatrix} F_0 & F_1 & F_2 & F_3 & \cdots & F_{m-1} \\ P_0 & 0 & 0 & 0 & \cdots & 0 \\ 0 & P_1 & 0 & 0 & \cdots & 0 \\ 0 & 0 & P_2 & 0 & \cdots & 0 \\ \vdots & \vdots & \vdots & \ddots & \vdots & \vdots \\ 0 & 0 & 0 & \ldots & P_{m-2} & 0 \end{bmatrix}. \tag{3.22}$$

The solution to the model is $\mathbf{p}(t) = L^t\mathbf{p}(0)$.

We now embed this representation of age structure into a discrete spatial model. Let $\mathbf{p}_i$, $i = 0, \ldots, n$, be the population vector in one of n spatial patches and let L_i denote the Leslie matrix for that patch. If $M_{ij} = M_{ji}$ is a matrix representing net migration between patch i and patch j, and $M_i = \sum_j M_{ij}$, then one can (albeit quite naively) use $M_i\mathbf{p}_i$ as the population in patch i after migration has occurred. The question remains: occurred during what length of time?

Assume $M_i(t)$ changes with time, with t in this case being the time at the start of the step. Explicit time dependence in the matrix is used to reflect the implicit time dependence of migration in the partial differential equation models due to the density dependence within diffusion and other mechanisms of movement. We are interested in the situation where migration occurs on faster time scales than aging. Let $\Delta t = 1/N$ for some integer N where we recall that the time steps in the Leslie matrix model have duration equal to 1. Our model with aging and migration becomes, for patch i,

$$\mathbf{p}_i(t+1) = L_i M_i(t+(N-1)\Delta t) \cdots M_i(t+2\Delta t) M_i(t+\Delta t) M_i(t)\mathbf{p}_i(t). \tag{3.23}$$

This matrix model falls short of corresponding to the discretization of the continuous models discussed in this paper, since it represents a situation where the ages of individuals remain fixed during the sequence of migration steps. The continuous analog of this matrix model is one where individuals move along characteristic curves that are step functions in the age-time plane rather than lines with slope one. A closer correspondence is obtained by decoupling fecundity and survivorship in the Leslie matrix and attaching a notion of age to each index a of the vector $\mathbf{p}_i$. Instead of a single matrix L_i, we have two $m \times m$ matrices

$$F_i(\alpha) = \begin{bmatrix} F_{0,i}(\alpha) & F_{1,i}(\alpha) & F_{2,i}(\alpha) & F_{3,i}(\alpha) & \cdots & F_{m-1,i}(\alpha) \\ 0 & 0 & 0 & 0 & \cdots & 0 \\ \vdots & \vdots & \vdots & \vdots & \ddots & \vdots \\ 0 & 0 & 0 & 0 & \cdots & 0 \end{bmatrix} \tag{3.24}$$

and

$$P_i = \begin{bmatrix} 0 & 0 & 0 & 0 & \cdots & 0 \\ P_{0,i} & 0 & 0 & 0 & \cdots & 0 \\ 0 & P_{1,i} & 0 & 0 & \cdots & 0 \\ 0 & 0 & P_{2,i} & 0 & \cdots & 0 \\ \vdots & \vdots & \vdots & \ddots & \vdots & \vdots \\ 0 & 0 & 0 & \ldots & P_{m-2,i} & 0 \end{bmatrix}, \tag{3.25}$$

where α is a value between 0 and 1 representing aging of a cohort during the sequence of substeps of length Δt that comprise a full time step of duration 1. For example, if at time t we have $\alpha = 0$, we can think of $p_{i,0}(t)$ consisting solely of newborns with age 0 within patch i, $p_{i,1}(t)$ consisting of individuals with age within the interval $(0,1]$, $p_{i,2}(t)$ consisting of individuals with age within the interval $(1,2]$, and so forth. The fecundity of $\mathbf{p}_i$ would be given by $F_i(0)$. After a time step of length Δt, we set $\alpha = \Delta t$ and get that the elements of $\mathbf{p}_i(t+\Delta t)$ consist of individuals with ages within $[0,\Delta t]$ for $a=0$, $(\Delta t, 1+\Delta t]$ for $a=1$, $(1+\Delta t, 2+\Delta t]$ for $a=2$, and so forth. The fecundity of $\mathbf{p}_i$ would now be given by $F_i(\Delta t)$. After two steps of length Δt, we set $\alpha = 2\Delta t$ and proceed as before. When α becomes one, we multiply by the survivorship matrix, reset α to zero, and repeat the process. The matrix model for this process is

$$\mathbf{p}_i(t+1) = P_i M_i(t+(N-1)\Delta t) F_i((N-1)\Delta t) \cdots$$
$$\cdots M_i(t+2\Delta t) F_i(2\Delta t) M_i(t+\Delta t) F_i(\Delta t) M_i(t) F_i(0) \mathbf{p}_i(t). \quad (3.26)$$

This model is most likely inappropriate for situations where one would typically use a linear matrix model—if the population dynamics are this complex, then a model allowing for explicit nonlinearities is probably the better approach. However, we have hopefully gained some insight into how the moving-grid Galerkin method decouples the age and time discretization while allowing age and time to advance together. We also note that the choice of a constant substep of Δt was chosen for illustrative purposes. The methods for the continuous models presented in this paper allow for nonuniform time steps.

3.3. *Time and space integration*

The moving-grid Galerkin method described in Section 3.1 reduces the age- and space-structured evolution equation described in Equations (2.7)–(2.11) to a system of parabolic equations. We describe a means of completing the discretizations in time and space by applying a backward-Euler method in time and center finite difference method in space to Equation (3.17). The original treatment in [3] used a more general Galerkin formulation in space.

Let $W_i^j(x_l)$ denote the approximate solution on the i-th age interval, at the j-th time step t^j and the l-th space node x_l. Define $\Delta t^j = t^j - t^{j-1}$ and $\Delta c_i^j = \Delta c_i(t^j)$. We assume that the time discretization is a refinement of the age discretization. Specifically, we require that $[-t^j, -t^{j-1}] \subseteq [c_i, c_{i+1}]$ for some i. We use the notation $\bar{\mu}_i^j(\varphi) = \bar{\mu}_i(x, t^j, \varphi)$. Set $W^j(c, x_l) = \sum_i W_i^j(x_l)\chi_{C_i(t^j)}(c)$, where χ is the characteristic function. The term P^j is obtained by integrating W^j. The birth function is $B_i^j(\varphi) = B_i(x, t^j, \varphi)$, if $[-t^j, -t^{j-1}] \subseteq [c_i, c_{i+1}]$, or $B_i^j(\varphi) = 0$, otherwise. The fully discrete method is defined by the system

$$\frac{W_i^j(x_l) - W_i^{j-1}(x_l)}{\Delta t^j} - \frac{J_{l+1/2} - J_{l-1/2}}{\frac{1}{2}(x_{l+1} - x_{l-1})} + \bar{\mu}_i^j(P^{j-1}) W_i^j(x_l)$$
$$= B_i^j(W^{j-1}(x_l)), \quad (3.27)$$

for every i, j, and l, except those corresponding to the initial condition or boundary conditions in space. The fluxes are given by

$$J_{l+1/2} = \frac{1}{2}\big[k(x_{l+1}, P^{j-1}(x_{l+1}))$$
$$+ k(x_l, P^{j-1}(x_l))\big] \frac{W_i^j(x_{l+1}) - W_i^j(x_l)}{x_{l+1} - x_l}, \quad (3.28)$$

$$J_{l-1/2} = \frac{1}{2}\big[k(x_l, P^{j-1}(x_l)) + k(x_{l-1}, P^{j-1}(x_{l-1}))\big]\frac{W_i^j(x_l) - W_i^j(x_{l-1})}{x_l - x_{l-1}}. \tag{3.29}$$

By lagging P and B at each time step, the discrete equations are coupled to each other only by values which are known before the step is taken. Thus, for each age group, we solve a single linear system that is independent from those of the other age groups. This allows for relatively easier parallelization of the computations.

3.4. *Convergence results*

In this section we summarize convergence results for the methods discussed above with the conditions on the general model defined by Equations (2.7)–(2.11) needed for the analyses. The example systems to follow will deviate from the general model and conditions without apparent loss of accuracy. The theoretical convergence results provide confidence in the computational framework and a sense of *terra firma.*

Using the lowest-order approximation space of piecewise-constant functions discussed in Section 3.1, one would expect first-order convergence in age. However, the moving-grid Galerkin method possess a superconvergence property: the computed constant value over each age interval is a second-order accurate approximation of the average value over the age interval, which is in turn a second-order accurate approximation of the midpoint value of the true solution. A continuous piecewise-linear function constructed from the midpoint values of the computed solution is thus a second-order accurate approximation of the true age distribution. Constructing a higher-order piecewise-polynomial approximation from a lower-order piecewise-polynomial approximation with a superconvergence property is referred to as "postprocessing." This superconvergence result is proved in [3] for Equations (2.7)–(2.11), for the case when we use a backward-Euler method in time and a Galerkin method in space, with the following stipulations:

Coercivity and Boundedness of k and μ: There exists constants C_0 and C_1 such that for $(x,p) \in \Omega \times \mathbb{R}$, k satisfies $0 < C_0 \le k(x,p) \le C_1$ and μ satisfies $0 < C_0 \le \mu(x,a,p) \le C_1$ for all a.

Lipschitz Conditions: The functions $k(x,p)$ and $\mu(x,a,p)$ are uniformly Lipschitz continuous with respect to p with Lipschitz constant K. The derivative $\frac{\partial}{\partial p}k(x,p)$ exists. The derivative $\frac{\partial}{\partial a}\mu(x,a,p)$ exists,

is uniformly bounded by C_1 as a function of all its arguments, and $\|\frac{\partial}{\partial a}\mu(x,\cdot,p)\|^2_{L^2(\mathbb{R}^+)} \le C_1$ uniformly as a function of x and p.
The birth condition, $\mathcal{B} : \mathbb{R} \times \left(L^1(\mathbb{R}^+) \cap L^2(\mathbb{R}^+)\right) \to \mathbb{R}^+$, satisfies the Lipschitz condition

$$|\mathcal{B}(x,\varphi(x,\cdot,t)) - \mathcal{B}(x,\psi(x,\cdot,t))| \qquad (3.30)$$
$$\le K\left(\left(1+\|\varphi\|_{L^1(\mathbb{R}^+)}\right)\left|\int_0^\infty (\varphi-\psi)\, da\right| + \|\varphi-\psi\|_{H^{-1}(\mathbb{R}^+)}\right),$$

and is bounded. Here, $H^{-1}(\mathbb{R}^+)$ is the dual to $H^1(\mathbb{R}^+)$.
An example of the birth condition is

$$\mathcal{B}(x,\varphi(x,\cdot,t)) = \int_0^\infty \beta(x,a,\Phi)\varphi(x,a,t)\, da, \qquad (3.31)$$

where $\beta \ge 0$ is the birth rate and Φ is the integral of φ with respect to age. This birth condition is used most often in the literature and represents a situation where an individual's fecundity, β, is dependent on its age, position in space, and the total population density at that position. The birth condition is satisfied if β is uniformly Lipschitz continuous as a function of Φ; if $\beta(x,a,\Phi)$, considered as a function of a, is in $H^1(\mathbb{R}^+)$, with H^1-norm bounded independently of x and Φ; and $\varphi \in L^1(\mathbb{R}^+) \cap L^2(\mathbb{R}^+)$, as a function of age.

Initial Condition: The initial condition, $u_0(x,a)$, is bounded and non-negative and there exists $\tilde{a}_{\max}$ such that $u_0(x,a) = 0$ for $a > \tilde{a}_{\max}$.

The model for which convergence of the moving-grid Galerkin method was shown, and the above conditions, were chosen for analytical and theoretical reasons. They provide insight into why and when we can expect the numerical methods to succeed or fail, but do not formally cover all situations in which the methods are effective. For example, the stipulation that the diffusivity $k(x,p)$ is bounded away from zero is violated in both example systems presented below. In practice this has not caused any degradation in convergence.

Convergence using a moving-grid Galerkin method with higher-order discontinuous piecewise-polynomial functions as the approximation space in age was shown in [5]. In the formulation of the methods and in the analysis, time was kept continuous so as to elucidate the decoupling of age and time discretizations even as the solution moves along characteristics.

The convergence result in [5] indicated that the methods using these higher-order spaces also possess a superconvergence property. Formally, only one additional order of convergence was obtained in age, although example systems have indicated the expected doubling of the order of convergence. For example, discontinuous piecewise-linear functions normally provide second-order convergence. However, according to our formal superconvergence result, we can expect third-order convergence via postprocessing. In practice, we are able to obtain fourth-order convergence by post-processing the discontinuous piecewise-linear functions to continuous piecewise-cubic functions. This is done by interpolating the cubic over two intervals through the quadrature points for two-point Gaussian quadrature.

Another convergence result of note is that for a step-doubling Galerkin method for parabolic problems [6], which is used in conjunction with the moving-grid Galerkin method to obtain the solutions in the example systems below. Step-doubling is a time integration method where a solution obtained by backward Euler over a time step is compared to a solution obtained by two half steps of backward Euler over the same interval. In addition to comparing the two solutions as a measure of truncation error, these two first-order correct approximations can be combined to form a second-order correct method.

In all the convergence results space has been discretized using a Galerkin finite element method for greater generality. We get second-order convergence in space by using the center finite differences described in Equation (3.27).

4. Example Systems

In this section we present two example systems that differ in many ways from the general model presented in Equations (2.7)–(2.11), yet whose model equations were solved effectively by the moving-grid Galerkin methods presented in [3, 5], in conjunction with other methods for time and space integration. The first of these, *Proteus mirabilis* swarm-colony development, differs from the general model in that age now represents time spent in a particular cell form and as a proxy for size. These changes come with attendant differences in "birth" and "death", and that the age- and space-structured partial difference equation is coupled to an ordinary difference equation at each point in space. The second system, tumor invasion, differs from the general model in that it contains a system of age- and space-structured equations coupled to reaction diffusion and ordinary differential equations. Moreover, the system contains a host of additional nonlinearities, in particular a haptotaxis term that accounts for movement of cells up a chemical gradient.

4.1. *Proteus mirabilis swarm-colony development*

When an inoculation of liquid medium containing *Proteus* cells is placed on an agar surface, after a period of local population growth, the bacteria begin to spread radially for another period of time before stopping and consolidating once again. This behavior repeats itself and, after the initial swarming and stopping phase, the velocity of the colony front becomes periodic. Spatially, this results in a bull's-eye pattern of terraces where each ring or terrace has the same width. Moreover, the total time spent forming a ring (which includes the time spent swarming and the time spent consolidating) does not change if the agar or nutrient (glucose) concentration is altered within wide bounds. The ring width and ratio of swarm time to consolidation time does change with changes in the agar or glucose concentrations. Changes in temperature, which influence metabolic rate, do alter the total cycle time. The invariance of the total cycle time to changes in the substrate, along with biological observation, rules out reaction to the chemical environment as an explanation for the regularity in *Proteus* swarm-colony development. A summary of the physiology of *Proteus* cells and the macroscopic colony behavior can be found in [4], and in more detail in [30, 38].

The spatial and temporal regularity seen in *Proteus* is an expression of their cell cycle and transition between two cell types, immotile "dividing" cells (referred to as "swimmer" cells in much the literature due to their motility in liquid media) and motile multinuclear filament "swarmer" cells. The population dynamics of swarmer cells is described by an age- and space-structured partial differential equation, which is coupled to an ordinary different equation at each point in space that describes the dynamics of the dividing cell population. The model in this section was originally presented in [4], which drew many of its main mechanisms from the models of Esipov and Shapiro [15] and Medvedev, Kaper, and Kopell [27].

The nondimensional model treated in [4] is as follows. The swarmer population $u(r, a, t)$ is distributed in space, represented by radius r, age a, and time t. The dividing-cell population $v(r, t)$ is described at each radius by an ordinary differential equation. Movement of swarmer cells is represented by a degenerate diffusion with diffusivity $D(p)$ that depends on the total swarmer biomass p. Cell differentiation from dividing-cell type to swarmer-cell type is given by a probability $\xi(v)$ that depends on the dividing cell density. Dedifferentiation from swarmer-cell type to dividing-cell type occurs at age $a_{\max}$. We assume the initial inoculation contains only dividing cells, and use no-flux boundary conditions. The model equations are

$$u_t + u_a = \frac{1}{r}\left(r(D(p)u)_r\right)_r, \quad 0 \le r < 1,\, 0 < a < a_{\max},\, t > 0, \quad (4.32a)$$

$$v_t = (1 - \xi(v))v + u(r, a_{\max}, t)e^{a_{\max}}, \quad 0 \le r \le 1,\ t > 0, \quad (4.32b)$$

$$p(r,t) = \int_{a_{\min}}^{a_{\max}} u\, e^a \, da, \quad 0 \le r \le 1,\ t \ge 0, \quad (4.32c)$$

with conditions

$$u(r,0,t) = \xi(v)v(r,t), \quad 0 \le r \le 1,\ t > 0, \quad (4.32d)$$

$$\left(D(p(1,t))u(1,a,t)\right)_r = 0, \quad 0 < a < a_{\max},\ t > 0, \quad (4.32e)$$

$$u(r,a,0) = 0, \quad 0 \le r \le 1,\ 0 < a < a_{\max}, \quad (4.32f)$$

and

$$v(r,0) = \begin{cases} v_h\left(2\left(\frac{r}{r_0}\right)^3 - 3\left(\frac{r}{r_0}\right)^2 + 1\right), & 0 \le r \le r_0, \\ 0, & r > r_0. \end{cases} \quad (4.32g)$$

Proteus move through a process of raft building that requires two things: sufficient maturity in swarmer cells to contribute to raft building ($a_{\min}$), and a sufficient biomass of mature cells to form the rafts ($p_{\min}$). The diffusivity has the form

$$D(r,t) = D_0 \max\{(p(r,t) - p_{\min}), 0\}. \quad (4.32h)$$

We use a differentiation function with a lag phase that is a C^1 piecewise cubic with support of length 2:

$$\xi(v) = \begin{cases} \xi_0\left(2|v - v_c|^3 - 3(v - v_c)^2 + 1\right), & |v - v_c| \le 1, \\ 0, & \text{otherwise,} \end{cases} \quad (4.32i)$$

The interval $[v_c - 1, v_c + 1]$ is the swarmer-cell production window.

Although this model differs in a number of ways from the general model in Equations (2.7)–(2.11) and the assumptions made for the convergence results, in particular the degenerate diffusion, it was successfully solved using the moving-grid Galerkin method in age to reduce the problem to a parabolic system of partial differential equations, which was then solved using a step-doubling Galerkin method in time and space. Details of the computation are in the manuscript [4]. The computational advantages of this combined methodology are illustrated by a plot of the time steps taken

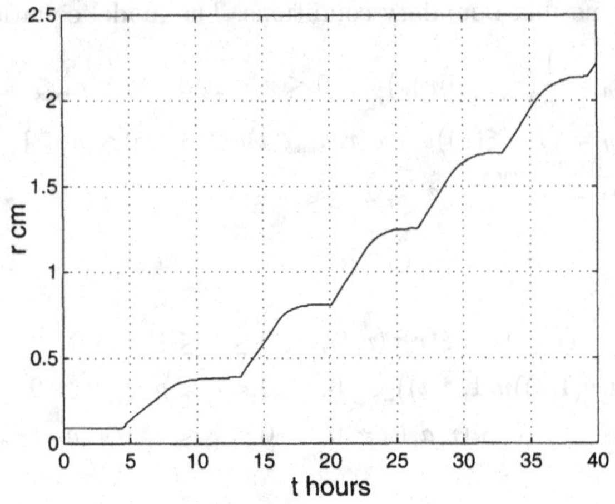

(a) Position of Colony Front

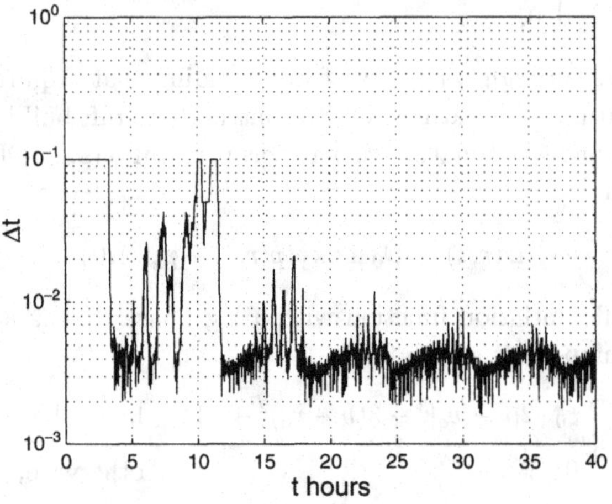

(b) Time Steps

Figure 3. (a) The position of the colony front showing the periodic behavior. (b) Time steps as a function of time. The size of the age intervals after the birth interval was 10^{-1}, significantly larger than what was needed for the time discretization. Note the connection between the oscillations in the size of the time steps and the state of the colony within its swarming and consolidation cycle.

during a simulation. Figure 3 shows that decoupling the time steps from the age step gives about a factor of 40 decrease in the number of age nodes, and that the ability to choose the time step adaptively gives roughly another factor of 10 increase in efficiency without loss of accuracy.

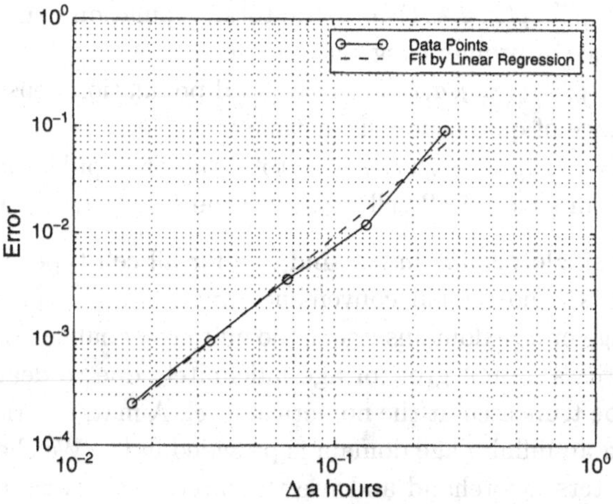

Figure 4. Convergence of the moving-grid Galerkin method for the *Proteus* model. The slope of the log-log plot of the age step versus the l^2 error gives the order of convergence. In this case the slope is approximately 2.07, which indicates second-order convergence.

Despite some significant differences between the *Proteus* model and the general model, the use of piecewise constants as the approximation space in age yields the expected second-order convergence in age, illustrated in Figure 4 as a log-log plot of the age step versus the l^2 error.

4.2. *Tumor invasion*

The following model was originally presented in [7] and has two explicit spatial dimensions denoted by $(x, y) \in \Omega$, whereas the *Proteus* model above assumed radial symmetry. The dependent variables of the model are

- $p(x, y, a, t)$ = density of proliferating tumor cells at position (x, y) and age a at time t.
- $q(x, y, a, t)$ = density of quiescent tumor cells at position (x, y) and age a at time t.
- $f(x, y, t)$ = surrounding tissue macromolecule (MM) density at position (x, y) at time t. It is assumed that these macromolecules are distributed heterogeneously in Ω, but immobile in Ω.
- $m(x, y, t)$ = matrix degradative enzyme (MDE) concentration at position (x, y) at time t. MDE is produced by the tumor cells and diffuses in Ω.
- $c(x, y, t)$ = oxygen concentration at position (x, y) at time t. Oxygen is produced by the extracellular MM, diffuses in Ω, and is consumed by the tumor cells.

- $P(x, y, t) = \int_0^{a_M} p(x, y, a, t)\, da =$ the total population density at a given point in space of proliferating cells at time t.
- $Q(x, y, t) = \int_0^{a_M} q(x, y, a, t)\, da =$ the total population density at a given point in space of quiescent cells at time t.
- $N(x, y, t) = P(x, y, t) + Q(x, y, t) =$ total tumor population density at a given point in space of all cell types at time t.

We use a single maximum age and size for all cell types and mutation classes. This is a notational convenience; models are often written with unbounded age or size domains under the quite reasonable assumption that biological entities do not grow or age indefinitely due to death. Thus, the domain can be truncated at the numerical level. A mathematical treatment on truncating an infinite age domain is provided in [3]. We choose to define these parameters beforehand and use the largest values we need to cover all cases.

$$\begin{aligned}
\frac{\partial}{\partial t} p(x, y, a, t) = &- \underbrace{\frac{\partial}{\partial a} p(x, y, a, t)}_{\text{cellaging}} \\
&+ \underbrace{D_p \nabla^2 p(x, y, a, t)}_{\text{diffusion}} - \underbrace{\chi \nabla \cdot \big(p(x, y, a, t) \nabla f(x, y, t)\big)}_{\text{haptotaxis}} \\
&- \underbrace{\rho(x, y, a, c) p(x, y, a, t)}_{\text{celldeathfrominsufficientoxygen}} - \underbrace{\theta(x, y, a, c) p(x, y, a, t)}_{\text{divisionwithsufficientoxygen}} \\
&- \underbrace{\sigma(x, y, a, c, N(x, y, t)) p(x, a, s, t)}_{\text{exit to quiescence}} \\
&+ \underbrace{\tau(x, y, a, c) q(x, y, a, t)}_{\text{entryfromquiescence}},
\end{aligned} \tag{4.33a}$$

$$\begin{aligned}
\frac{\partial}{\partial t} q(x, y, a, t) = &- \underbrace{\frac{\partial}{\partial a} q(x, y, a, t)}_{\text{cellaging}} - \underbrace{\nu(x, y, a, c) q(x, y, a, t)}_{\text{celldeathfrominsufficientoxygen}} \\
&+ \underbrace{\sigma(x, y, a, c, N(x, y, t)) p(x, y, a, t)}_{\text{entryfromproliferation}} \\
&- \underbrace{\tau(x, y, a, c) q(x, y, a, t)}_{\text{exittoproliferation}},
\end{aligned} \tag{4.33b}$$

with age-boundary conditions

$$\underbrace{p(x, y, 0, t)}_{\text{newborncells}} = 2 \underbrace{\int_0^{a_M} \theta(x, y, a, c) p(x, y, a, t)\, da}_{\text{division rate}}. \tag{4.33c}$$

The equations governing tissue macromolecule, matrix degradative enzyme, and oxygen densities are those originally used in [1]:

$$\frac{\partial}{\partial t}f(x,y,t) = \underbrace{-\ \delta m(x,y,t)f(x,y,t)}_{\text{degradation}}, \tag{4.33d}$$

$$\begin{aligned}\frac{\partial}{\partial t}m(x,y,t) &= \underbrace{D_m\nabla^2 m(x,y,t)}_{\text{diffusion}} \\ &\quad + \underbrace{\mu P(x,y,t) + \omega Q(x,y,t)}_{\text{production}} - \underbrace{\lambda m(x,y,t)}_{\text{decay}},\end{aligned} \tag{4.33e}$$

$$\begin{aligned}\frac{\partial}{\partial t}c(x,y,t) &= \underbrace{D_c\nabla^2 c(x,y,t)}_{\text{diffusion}} + \underbrace{\beta f(x,y,t)}_{\text{production}} \\ &\quad \underbrace{-\ \gamma P(x,y,t) - \eta Q(x,y,t)}_{\text{uptake}} - \underbrace{\alpha c(x,y,t)}_{\text{decay}}.\end{aligned} \tag{4.33f}$$

All equations are combined with initial conditions and zero flux boundary conditions on an (x, y)-rectangle Ω. An illustrative example computation of the system defined by Equations (4.33a)–(4.33f) was presented in [7] . Animations can be found online at http://faculty.smu.edu/ayati/cancer.html.

As with the *Proteus*, this model differs in a number of ways from the general model in Equations (2.7)–(2.11) and the assumptions made for the convergence results. In addition to the diffusion terms being allowed to become degenerate, the tumor model contains a haptotaxis term not considered in the general model. Nonetheless, it was successfully solved using the moving-grid Galerkin method in age to reduce the problem to a parabolic system of partial differential equations. The parabolic system was then solved using a step-doubling alternating-direction implicit (ADI) method in time and space. The ADI method consists of solving the system of equations first along the x-direction and zero-order terms, and then along the y-direction. This has the effect of reducing a wide-banded matrix that would have been obtained by a direct discretization to a series of smaller block-tridiagonal matrices. A discussion of the step-doubling ADI method and other details of the computation are in [4].

5. Conclusions

In this paper we have attempted to clarify for an audience of mathematical biologists, rather than numerical analysts, one approach to the modeling and simulation of age- and space-structured biological systems. We began with a general model through which we presented the main aspect of the computational methodology: the moving-grid Galerkin method used to decouple the age and time discretizations while still having age and time advance together. We then put it in the context of some other methods for the time and space integration before discussing the numerical analysis literature and presenting some example systems.

One observation has been that, in many systems, age-structure is a contributing factor to the behavior of interest, but has been omitted in models due to the costs associated with the additional complexity. A major goal of the development of the numerical methods for age- and space-structured systems was to encourage the use of age structure when appropriate. This includes situations where one my settle upon using a coarse stage-structured model. In these cases, the moving-grid Galerkin method applied to an age-continuous model may yield a much more accurate representation of the biological system by using an age discretization with as few as a dozen age intervals.

As shown in the example systems, where age was used as a proxy for cell length in the *Proteus* model and position within the cell cycle in the tumor invasion model, age can often be used in place of other forms of physiological structure when a clear relationship exists between the physiological trait of interest and some sense of time as measured by an age variable.

References

[1] Alexander R. A. Anderson. A hybrid mathematical model of solid tumour invasion: The importance of cell adhesion. *Math. Med. Biol. IMA Journal*, **22** (2005) 163–186.

[2] D. G. Aronson. The role of diffusion in mathematical population biology: Skellam revisited. In V. Capasso, E. Grosso, and S. L. Paveri-Fontana, editors, *Mathematics in Biology and Medicine*, volume 57 of *Lecture Notes in Biomathematics*, pages 2–6. Springer-Verlag, Berlin, 1985.

[3] Bruce P. Ayati. A variable time step method for an age-dependent population model with nonlinear diffusion. *SIAM J. Numer. Anal.*, **37** (5) (2000) 1571–1589.

[4] Bruce P. Ayati. A structured-population model of *Proteus mirabilis* swarm-colony development. *J. Math. Biol.*, **52** (1) (2006) 93–114.

[5] Bruce P. Ayati and Todd F. Dupont. Galerkin methods in age and space for a population model with nonlinear diffusion. *SIAM J. Numer. Anal.*, **40**(3) (2002) 1064–1076.

[6] Bruce P. Ayati and Todd F. Dupont. Convergence of a step-doubling Galerkin method for parabolic problems. *Math. Comp.*, **74** (251) (2005) 1053–1065.

[7] Bruce P. Ayati, Glenn F. Webb and Alexander R. A. Anderson. Computational methods and results for structured multiscale models of tumor invasion. *Multiscale Model. Simul. SIAM*, **5**(1) (2006) 1–20.

[8] S. Busenberg and M. Iannelli. A class of nonlinear diffusion problems in age-dependent population dynamics. *Nonlin. Anal. Th. Meth. Appl.*, **7** (1983) 501–529.

[9] J. M. Cushing. *An Introduction to Structured Population Dynamics*, volume 71 of *CBMS-NSF Regional Conference Series in Applied Mathematics.* SIAM, Philadelphia, 1998.

[10] A. M. de Roos. Numerical methods for structured population models: The escalator boxcar train. *Num. Meth. Part. Diff. Eqns.*, **4** (1989) 173–195.

[11] A. M. de Roos. A gentle introduction to physiologically structured population models. In S. Tuljapurkar and H. Caswell, editors, *Structured-population Models in Marine, Terrestrial, and Freshwater Systems*, volume 18 of *Population and Community Biology Series*, Chapter 5, pages 119–204. Chapman & Hall, New York, 1997.

[12] A. M. de Roos and J. A. J. Metz. Towards a numerical analysis of the escalator boxcar train. In J. A. Goldstein, F. Kappel, and W. Schappacher, editors, *Differential Equations with Applications in Biology, Physics and Engineering*, volume 133 of *Lecture Notes in Pure and Applied Mathematics*, pages 91–113. Marcel Dekker, New York, 1991.

[13] G. di Blasio. Non-linear age-dependent population diffusion. *J. Math. Bio.*, **8** (1979) 265–284.

[14] G. di Blasio and L. Lamberti. An initial-boundary problem for age-dependent population diffusion. *SIAM J. Appl. Math.*, **35** (1978) 593–615.

[15] Sergei E. Esipov and J. A. Shapiro. Kinetic model of *Proteus mirabilis* swarm colony development. *J. Math. Biol.*, **36** (1998) 249–268.

[16] M. E. Gurtin. A system of equations for age-dependent population diffusion. *J. Theor. Biol.*, **40** (1973) 389–392.

[17] M. E. Gurtin and R. C. MacCamy. Non-linear age-dependent population dynamics. *Arch. Rat. Mech. Anal.*, **54** (1974) 281–300.

[18] M. E. Gurtin and R. C. MacCamy. Diffusion models for age-structured populations. *Math. Biosci.*, **54** (1981) 49–59.

[19] C. Huang. An age-dependent population model with nonlinear diffusion in $\mathbb{R}^n$. *Quart. Appl. Math.*, **52** (1994) 377–398.

[20] Mark Kot. *Elements of Mathematical Ecology.* Cambridge University Press, Cambridge, United Kingdom, 2001.

[21] K. Kubo and M. Langlais. Periodic solutions for a population dynamics problem with age-dependence and spatial structure. *J. Math. Biol.*, **29** (1991) 363–378.

[22] M. Langlais. A nonlinear problem in age-dependent population diffusion. *SIAM J. Math. Anal.*, **16** (1985) 510–529.

[23] M. Langlais. Large time behavior in a nonlinear age-dependent population dynamics problem with spatial diffusion. *J. Math. Biol.*, **26** (1988) 319–346.

[24] R. C. MacCamy. A population model with nonlinear diffusion. *J. Diff. Eqns.*, **39** (1981) 52–72.

[25] P. Marcati. Asymptotic behavior in age-dependent population dynamics with hereditary renewal law. *SIAM J. Math. Anal.*, **12** (1981) 904–916.

[26] A. G. McKendrick. Applications of mathematics to medical problems. *Proc. Edin. Math. Soc.*, **44** (1926) 98–130.

[27] Georgiy S. Medvedev, Tasso J. Kaper and Nancy Kopell. A reaction-diffusion equation with periodic front dynamics. *SIAM J. Appl. Math.*, **60** (5) (2000) 1601–1638.

[28] Thomas Nagylaki. The diffusion model for migration and selection. In A. Hastings, editor, *Some Mathematical Questions in Biology: Models in Population Biology*, volume 20 of *Lectures on Mathematics in the Life Sciences*, pages 55–75. American Mathematical Society, Providence, R.I., 1989.

[29] Hans G. Othmer and Angela Stevens. Aggregation, blowup, and collapse: The ABC's of taxis in reinforced random walks. *SIAM J. Appl. Math.*, **57** (4): (1997) 1044–1081.

[30] O. Rauprich, M. Matsushita, K. Weijer, F. Siegert, S. E. Esipov and J. A. Shapiro. Periodic phenomena in *Proteus mirabilis* swarm colony development. *J. Bacteriol.*, **178** (1996) 6525–6538.

[31] M. Rotenberg. Theory of population transport. *J. Theor. Biol.*, **37** (1972) 291–305.

[32] F. R. Sharpe and A. J. Lotka. A problem in age-distribution. *Philos. Mag.*, **21** (6) (1911) 435–438.

[33] J. G. Skellam. Random dispersal in theoretical populations. *Biometrika*, **38** (1951) 196–218.

[34] J. W. Thomas. *Numerical Partial Differential Equations: Finite Difference Methods*, volume 22 of *Texts in Applied Mathematics*. Springer-Verlag, New York, 1995.

[35] Peter Turchin. Population consequences of aggregative movement. *J. Animal Ecology*, **58** (1989) 75–100.

[36] G. F. Webb. *Theory of Nonlinear Age-dependent Population Dynamics*, volume 89 of *Pure and Applied Mathematics*. Marcel Dekker, New York, 1985.

[37] Glenn F. Webb. An age-dependent epidemic model with spatial diffusion. *Arch. Rational Mech. Anal.*, **75** (1) (1980/81) 91–102.

[38] Fred D. Williams and Robert H. Schwarzhoff. Nature of the swarming phenomenon in *Proteus*. *Ann. Rev. Microbiol.*, **32** (1978) 101–122.

NUTRIENT-PLANKTON INTERACTION WITH A TOXIN IN A VARIABLE INPUT NUTRIENT ENVIRONMENT

SOPHIA R.-J. JANG

Department of Mathematics, University of Louisiana at Lafayette, Lafayette, LA 70504-1010

JAMES BAGLAMA

Department of Mathematics, University of Rhode Island, Kingston, RI 02881-0816

A simple model of phytoplankton-zooplankton interaction with a periodic input nutrient is presented. The model is then used to study a nutrient-plankton interaction with a toxic substance that inhibits the growth rate of phytoplankton. The effects of the toxin upon the existence, magnitude, and stability of the periodic solutions are discussed.

Keywords: plankton populations, toxin, inhibitor, uniform persistence.

1. Introduction

Phytoplankton are microscopic plants that live in the ocean, sea or lake. Through photosynthesis, phytoplankton are responsible for much of the oxygen present in the Earth's atmosphere. They convert inorganic materials into new organic compounds by the process of photosynthesis [14]. Hence the stocks of these tiny planktonic algae play a significant role for marine reserves and fishery management. In terms of numbers, the most important groups of phytoplankton are the diatoms, cyanobacteria and dinoflagellates, although many other groups of algae are also very populated.

Pollution of freshwater and marine systems by anthropogenic sources has become a concern over the last several decades. Organic (e.g. triazine herbicides) [1, 16–18, 25, 26] and inorganic compounds (e.g. heavy metals) [6, 13, 17–19, 23, 24] both may have harmful effects to the organsims. For example, samples taken from the inner harbor of the Waukegan area, located in Lake County, Illinois, on the west shore of Lake Michigan, have shown that photosynthesis of the green algae *Selenastrum capricornutum* is inhibited due to pollutants originating from industrial and recreational sources.

This study investigates the possible effects of toxic substances upon nutrient-phytoplankton-zooplankton interaction. In the early 1980s, Hansen and Hubbell [10] used antibiotic, nalidixic acid, to examine competition of two strains of *E. coli.* One strain was sensitive and the other was resistant to the inhibitor. This resistance of the population is due to a chromosomal mutation and it does not result in detoxification of the antibiotic. However, resistance by bacteria to antibiotics and heavy metals frequently comes from the acquisition of an extrachromosomal element that encodes an enzyme, which converts the inhibitor into a less toxic form. This reduces the intracellular toxic concentration and enables the survival of bacteria that produce the enzyme. It also results in a significant reduction of the inhibitor in the environment [2, 15]. This biological phenomenon motivates our plankton-toxin model proposed here.

The model consist of a single limiting nutrient, two plankton populations and the inhibitor, where the inhibitor may include agents such as pesticides or heavy metals. The phytoplankton feeds on nutrient and zooplankton grazes on phytoplankton. The zooplankton absorbs the inhibitor without effect, while phytoplankton's uptake rate and consequently its growth rate are inhibited due to the presence of an external inhibitor. Although our models are simplified systems, it is a first step in understanding complex interaction between the first two trophic levels and pollution. More complicated plankton systems such as multiple nutrients can be found in Grover [8].

We first propose a simple plankton model with a periodic input nutrient concentration and summarize its dynamical consequences. We then examine the effect of toxin upon the existence, magnitude, and stability of the periodic solutions. Criteria for coexistence of both plankton populations are also discussed. However, comparisons between more complex dynamical behavior will only be numerically simulated. A base nutrient-plankton model of closed ecosystem is presented in the next section. Section 3 studies the model when phytoplankton is inhibited by the toxin. Numerical examples will be provided to illustrate complexity of the interaction. The final section provides a brief summary and discussion.

2. The Nutrient-Phytoplankton-Zooplankton Model

In this section we shall introduce a base model which will be used to study the effects of toxin upon plankton interaction later. For simplicity, it is assumed that the organisms and the nutrient are uniformly distributed over the space. Let $N(t)$, $P(t)$, and $Z(t)$ denote the nutrient concentration, the phytoplankton population, and zooplankton population at time t, respectively. For convenience, the two plankton levels are modeled in terms of nutrient content and therefore their units are nitrogen or nitrate per unit volume. We let δ and ϵ denote the per capita natural death rate of phytoplankton and zooplankton respectively. The phytoplankton's nutrient uptake rate is denoted by f, while g is the zooplankton's grazing rate. Since plankton populations are measured in terms of nutrient concentration, f and g are functions of nutrient concentration. Both functions have the standard monotonic assumptions as the classical Michaelis-Menton kinetics, Ivlev, and Holling type III functional responses given below:

$$(\mathrm{H1}) f, g \in C^1[0,\infty), \quad f(0) = g(0) = 0,$$
$$f'(x), g'(x) > 0 \quad \text{for } x \geq 0 \quad \text{and} \quad \lim_{x\to\infty} f(x) = \lim_{x\to\infty} g(x) = 1.$$

Parameter m is the maximal nutrient uptake rate of phytoplankton and c denotes the maximal zooplankton ingestion rate, where β and α are the fraction of zooplankton grazing conversion and phytoplankton nutrient conversion, respectively. In natural nutrient-plankton systems, waters flowing into the system bring input of fluxes of nutrients and outflows also carry out nutrients [3, 7, 20, 21]. Unlike the study in [11], we assume that the input nutrient concentration $N^*(t)$ is varied periodically around N^0 with $N^0(t) = N^0 + ae(t)$, where $N^0 > 0$, $0 < a < N^0$, and $e(t)$ is τ-periodic with mean value zero and $|e(t)| \leq 1$ for $t \geq 0$. It is assumed that the rate of waters flowing in and out of the system is a constant and denoted by D. Both plankton populations are also assumed to be flowing out of the system with the same constant washout rate D.

Nutrients are consumed by the phytoplankton, which in turn is grazed upon by the herbivorous zooplankton. Consequently, there are minus terms $-mf(N)P$ and $-cg(P)Z$ in the equations for $\dot{N}$ and $\dot{P}$, respectively. For simplicity, we assume that the system under study is closed and hence there are positive feedback terms δP, ϵZ, $(1-\alpha)mf(N)P$, and $(1-\beta)cg(P)Z$ appeared in the equation for $\dot{N}$. Our model with the above biological assumptions can be written as the following three dimensional nonautonomous ordinary differential equations.

$$\begin{aligned}
\dot{N} &= D(N^0(t) - N) - mf(N)P + \delta P + \epsilon Z + (1-\beta)cg(P)Z \\
&\quad + (1-\alpha)mf(N)P \\
\dot{P} &= [\alpha mf(N) - \delta - D]P - cg(P)Z \\
\dot{Z} &= [\beta cg(P) - \epsilon - D]Z \\
&\quad N(0), P(0), Z(0) \geq 0,
\end{aligned} \tag{2.1}$$

where $0 < \alpha, \beta \leq 1$, and $D, N^0, m, c, \epsilon, \delta > 0$.

Notice the scalar periodic equation

$$\dot{N} = D(N^0(t) - N) \qquad (2.2)$$
$$N(0) \geq 0$$

has a unique positive τ-periodic solution

$$N^*(t) = \frac{De^{-Dt}}{e^{Dt} - 1} \int_t^{t+\tau} e^{Dr}[N^0 + ae(t)]dr$$

and solutions $N(t)$ of (2.2) can be written as $N(t) = N^*(t) + (N(0) - N^*(0))e^{-Dt}$ for all t. Hence solutions are asymptotic to the periodic solution $N^*(t)$. Since $\dot{N}|_{N=0} \geq DN^0(t) \geq 0$, $\dot{P}|_{P=0} = \dot{Z}|_{Z=0} = 0$, solutions of (2.1) remain nonnegative.

Let $U = N^*(t) - N - P - Z$. Then $\dot{U} = -DU$ and hence solutions of (2.1) are bounded. Moreover, system (2.1) can be rewritten as

$$\begin{aligned} \dot{U} &= -DU \\ \dot{P} &= [\alpha m f(N^*(t) - U - P - Z) - \delta - D]P - cg(P)Z \\ \dot{Z} &= [\beta c g(P) - \epsilon - D]Z. \end{aligned} \qquad (2.3)$$

Since the ω-limit set of (2.3) lies on the set $U = 0$, (2.1) has the following limiting system:

$$\begin{aligned} \dot{P} &= [\alpha m f(N^*(t) - P - Z) - \delta - D]P - cg(P)Z \\ \dot{Z} &= [\beta c g(P) - \epsilon - D]Z \\ & P(0), Z(0) \geq 0, P(0) + Z(0) \leq N^*(0). \end{aligned} \qquad (2.4)$$

As $N(t) + P(t) + Z(t) = N^*(t)$ for $t \geq 0$ on the ω-limit set and solutions of (2.1) remain nonnegative, we see that $P(t) + Z(t) \leq N^*(t)$ for $t \geq 0$, i.e., system (2.4) is well-defined.

Let

$$\Gamma = \{(P, Z) \in R_+^2 : P + Z \leq N^*(0)\}.$$

System (2.4) has a trivial solution $(0,0)$ for all parameter values. The Jacobian derivative of the Poincaré map induced by system (2.4) evaluated at $(0,0)$ is given by $\Phi_0(t)$, where $\Phi_0(t)$ is the fundamental matrix solution of $\dot{X} = J_0 X$ with

$$J_0 = \begin{pmatrix} \alpha m f(N^*(t)) - \delta - D & 0 \\ 0 & -\epsilon - D \end{pmatrix}. \qquad (2.5)$$

Let

$$\sigma_0 = \frac{1}{\tau} \int_0^{\tau} [\alpha m f(N^*(t)) - \delta - D]dt.$$

Theorem 2.1. *If $\sigma_0 < 0$, then solutions of (2.4) satisfy* $\lim_{t\to\infty} P(t) = \lim_{t\to\infty} Z(t) = 0$.

Proof. We may assume $P(0) > 0$. Since $\dot{P} \le [\alpha m f(N^*(t)) - \delta - D]P$ for $t \ge 0$, consider the following equation

$$\dot{x} = [\alpha m f(N^*(t)) - \delta - D]x$$

with $x(0) = P(0)$. The solution can be written explicitly as

$$\begin{aligned} x(t) &= x(0)e^{\int_0^t [\alpha m f(N^*(r))-\delta-D]dr} \\ &= x(0)e^{\int_0^{t_0} [\alpha m f(N^*(r))-\delta-D]dr} \\ &\quad \times e^{\int_{t_0}^{n\tau+t_0} [\alpha m f(N^*(r))-\delta-D]dr} \\ &= x(0)e^{\int_0^{t_0} [\alpha m f(N^*(r))-\delta-D]dr} \\ &\quad \times e^{\int_0^{n\tau} [\alpha m f(N^*(r))-\delta-D]dr} \end{aligned}$$

for some $0 \le t_0 < \tau$ and $n > 0$, where t_0 and n depend on t. Notice $t \to \infty$ if and only if $n \to \infty$. Hence $\lim_{t\to\infty} x(t) = 0$ as $\sigma_0 < 0$. As a result, $\lim_{t\to\infty} P(t) = 0$. Therefore for any $\eta > 0$, there exists $t_1 > 0$ such that $P(t) < \eta$ for $t \ge t_1$. We choose $\eta > 0$ such that $\beta c g(\eta) < \epsilon + D$. It follows from the equation for $\dot{Z}$ in (2.4) that $\lim_{t\to\infty} Z(t) = 0$ and this completes the proof. □

Suppose now $\sigma_0 > 0$. Consider the linear periodic system

$$\dot{X} = J_0 X \tag{2.6}$$

where J_0 is given in (2.5) and X is a row vector. Let $\Phi(t)$ be the fundamental matrix solution of the linear system (2.6) with $\Phi(0) = I$, the identity matrix. Then the Floquet multipliers of $(0, 0)$ are the eigenvalues of $\Phi(\tau)$ [5]. Since

$$\Phi(\tau) = \begin{pmatrix} e^{\int_0^\tau [\alpha m f(N^*(t))-\delta-D]dt} & 0 \\ 0 & e^{-(\epsilon+D)\tau} \end{pmatrix} \tag{2.7}$$

and $\sigma_0 > 0$, we see that $(0, 0)$ is unstable.

Theorem 2.2. *Suppose $\sigma_0 > 0$. Then (2.4) has a unique τ-periodic solution $(\bar{P}(t), 0)$ with $\bar{P}(t) > 0$. Moreover, solutions of (2.4) with $P(0) > 0$ and $Z(0) = 0$ converge to $(\bar{P}(t), 0)$ asymptotically.*

Proof. Since $Z(t) = 0$ for $t > 0$ if $Z(0) = 0$, we consider the following equation

$$\begin{aligned} \dot{P} &= [\alpha m f(N^*(t) - P) - \delta - D]P \\ 0 &\le P(0) \le N^*(0). \end{aligned} \tag{2.8}$$

Let $T_0 : [0, N^*(0)] \to [0, N^*(0)]$ denote the Poincaré map induced by equation (2.8), i.e., $T_0(P_0) = P(\tau, P_0)$, where $P(t, P_0)$ is the solution of (2.8) with $P(0) = P_0$.

Notice $T_0(0) = 0$, $T_0(N^*(0)) < N^*(0)$ and $\dot{T}_0 = \frac{\partial(\tau, P_0)}{\partial P_0} = v(\tau)$, where $v(t)$ satisfies

$$\begin{aligned} \dot{P} &= [\alpha m f(N^*(t) - P) - \delta - D - \alpha m f'(N^*(t) - P)P]v \\ v(0) &= 1. \end{aligned}$$

Therefore, $\dot{T}_0 > 0$, and in particular when $P_0 = 0$ we have

$$v(\tau) = e^{\int_0^\tau [\alpha m f(N^*(t)) - \delta - D)]dt}.$$

Thus $\dot{T}_0(0) > 1$, and the map T_0 has a unique positive fixed point $\bar{p}$, $\bar{p} < N^*(0)$, which corresponds to a unique positive τ-periodic solution $\bar{P}(t)$ for equation (2.8). Since T_0 is monotone increasing, it can be easily shown that $\lim_{n\to\infty} T_0^n(p) = \bar{p}$ for $0 < p \le N^*(0)$. Consequently, solutions of (2.8) with $P(0) > 0$ satisfy $\lim_{t\to\infty}(P(t) - \bar{P}(t)) = 0$. The proof is then complete. □

Let

$$\sigma_1 = \frac{1}{\tau}\int_0^\tau [\beta c g(\bar{P}(t)) - \epsilon - D]dt.$$

Theorem 2.3. *Let $\sigma_0 > 0$ and $\sigma_1 < 0$. Then solutions of (2.4) with $P(0) > 0$ satisfy $\lim_{t\to\infty}(P(t) - \bar{P}(t)) = \lim_{t\to\infty} Z(t) = 0$.*

Proof. We claim that $\lim_{t\to\infty} Z(t) = 0$. Since $\dot{P} \le [\alpha m f(N^*(t) - P) - \delta - D]P$ for all $t \ge 0$, consider the following equation

$$\begin{aligned} \dot{x} &= [\alpha m f(N^*(t) - x) - \delta - D]x \\ x(0) &= P(0). \end{aligned} \tag{2.9}$$

Observe that $P(t) \le x(t)$ for $t \ge 0$. Since $x(t) \to \bar{P}(t)$ as $t \to \infty$ by Theorem 2.2, $\liminf_{t\to\infty}(x(t) - \bar{P}(t)) = 0$. Hence for any $\eta > 0$ given, there exists $t_0 > 0$ such that $x(t) \le \bar{P}(t) + \eta$ for $t \ge t_0$. As a result, $P(t) \le \bar{P}(t) + \eta$ for $t \ge t_0$. By our assumption we can choose $\eta > 0$ such that

$$\int_0^\tau [\beta c g(\bar{P}(t) + \eta) - \epsilon - D]dt < 0.$$

Consequently, $\dot{Z} \le [\beta c g(\bar{P}(t) + \eta) - \epsilon - D]Z$ for $t \ge t_0$ implies $\lim_{t\to\infty} Z(t) = 0$.

It remains to show that $\lim_{t\to\infty}(\bar{P}(t) - P(t)) = 0$. Consider the Poincaré map T induced by system (2.4), $T(P_0, Z_0) = (P(\tau), Z(\tau))$, where $(P(t), Z(t))$ is the solution of (2.4) with initial condition (P_0, Z_0). Since $\lim_{t\to\infty} Z(t) = 0$, $\lim_{n\to\infty} T^n(P_0, Z_0)$ lies on the P-axis. Moreover, $T^n(P_0, 0) = (T_0^n P_0, 0)$, where T_0 is the Poincaré map associated with equation (2.8). Since T_0 has a unique positive fixed point $\bar{p}$ which is moreover globally asymptotically stable for T_0 in $(0, N^*(0)]$, it follows that $T^n(P_0, 0)$ converges to the fixed point $(\bar{p}, 0)$. Therefore the periodic solution $(\bar{P}(t), 0)$ is globally asymptotically stable. □

Suppose now $\sigma_0, \sigma_1 > 0$. Then the floquet multipliers are the eigenvalues of $\Phi_1(\tau)$, where $\Phi_1(t)$ is the fundamental matrix solution of $\dot{X} = J_1 X$, where

$$J_1 = \begin{pmatrix} J_{11} & -\alpha m f'(N^*(t) - \bar{P}(t))\bar{P}(t) - cg(\bar{P}(t)) \\ 0 & \beta c g(\bar{P}(t)) - \epsilon - D \end{pmatrix}, \tag{2.10}$$

and

$$J_{11} = \alpha m f(N^*(t) - \bar{P}(t)) - \delta - D - \alpha m f'(N^*(t) - \bar{P}(t))\bar{P}(t).$$

It follows that the periodic solution $(\bar{P}(t), 0)$ is unstable as $\sigma_1 > 0$. Similar to the arguments used in [11] we can prove that both populations can coexist by using the concepts of uniform persistence.

Theorem 2.4. *If $\sigma_0, \sigma_1 > 0$, then system (2.1) is uniformly persistent.*

Proof. We first apply Theorem 3.1 of Butler and Waltman [4] to show uniform persistence of the limiting system (2.4). Let $\mathcal{F}$ be the flow generated by system (2.4) and $\partial\mathcal{F}$ be $\mathcal{F}$ restricted to the boundary Γ. We need to verify that $\partial\mathcal{F}$ is isolated and acyclic. Let $M_0 = \{(0,0)\}$ and $M_1 = \{(\bar{P}(t), 0) : 0 \le t \le \tau\}$. Then the invariant set of $\partial\mathcal{F}$ is $\{M_0, M_1\}$. It is clear that $\partial\mathcal{F}$ is acyclic as M_0 and M_1 are globally attracting on the positive Z-axis and P-axis respectively and thus no subset of $\{M_0, M_1\}$ can form a cycle.

It remains to prove that each M_i is isolated for $\partial\mathcal{F}$ and for $\mathcal{F}$ respectively, for $i = 0, 1$. We only verify that M_0 is isolated for $\mathcal{F}$ as the remaining assertion can be argued similarly. Let $c_0 = \max_{0 \le P \le N^*(0)} g'(P)$. By our assumption we can choose $\rho > 0$ such that

$$\frac{1}{\tau}\int_0^\tau [\alpha m f(N^*(t) - \rho) - \delta - D - cc_0\rho]dt > 0. \tag{2.11}$$

Let $\mathcal{N} = \{(P, Z) \in \Gamma : d((P, Z), M_0) < \rho\}$, where d is the Euclidean metric on R^2. We show that $\mathcal{N}$ is an isolated neighborhood of M_0 in Γ.

If this is not true, then there exists an invariant set V in Γ such that $M_0 \subset V \subset \mathcal{N}$ and $V \backslash M_0 \neq \emptyset$. Notice we can find $P(0), Z(0) > 0$ such that $(P(0), Z(0)) \in V \backslash M_0$. On the other hand, $V \subset \mathcal{N}$ implies

$$\begin{aligned}\frac{\dot{P}}{P} &= \alpha m f(N^*(t) - P - Z) - \delta - D - \frac{cg(P)}{P} Z \\ &\geq \alpha m f(N^*(t) - \rho) - \delta - D - cc_0\rho.\end{aligned}$$

Hence

$$P(t) \geq P(0) e^{\int_0^t [\alpha m f(N^*(s) - \rho) - \delta - D - c_0 c\rho] ds}$$

and we have $\lim_{t\to\infty} P(t) = \infty$ by inequality (2.11). This is impossible as solutions of (2.4) are bounded. Therefore M_0 must be isolated in $\partial\mathcal{F}$. Furthermore, let $\mathring{\Gamma}$ denote the interior of Γ and $W_i^+(M_i)$ be the stable manifold of M_i, $i = 0, 1$. It follows from the Floquet multipliers of M_i that $W^+(M_i)$ is disjoint from $\mathring{\Gamma}$ for $i = 0, 1$. Therefore (2.4) is uniformly persistent by [4].

We now rewrite system (2.4) as $\dot{Y} = F(Y, t)$ and system (2.3) as $\dot{X} = F(X, t) + R(X, t)$. Therefore there exists $C = D \max_{0 \leq t \leq \tau} N*(t)$ such that $|R(X, t)| \leq Ce^{-Dt}$ for $t \geq 0$ for all solution $X(t)$ of system (2.3). As a result, Lemma A.4 of Hale and Somolinos [9] implies that the asymptotic behavior of (2.3) and (2.4) are the same. Since systems (2.1) and (2.3) are equivalent, we can conclude that system (2.1) is uniformly persistent. □

In summary, if the average maximal growth rate $\frac{1}{\tau}\int_0^\tau [\alpha m f(N^*(t)) - \delta - D]dt$ of phytoplankton is less than the total removal rate $\delta + D$, then phytoplankton population goes extinct and so does the zooplankton. If the average maximal growth rate of phytoplankton exceeds its total removal rate then the phytoplankton population can stabilize in a positive periodic solution fashion, $\bar{P}(t)$, in the absence of zooplankton. Consequently, zooplankton population becomes extinct if its average maximal growth rate $\frac{1}{\tau}\int_0^\tau [\beta c g(\bar{P}(t)) - \epsilon - D]dt$ when phytoplankton is stabilized, is less than its total removal rate $\epsilon + D$, and both populations can coexist if these average maximal growth rates are greater than the total removal rates.

3. A Nutrient-Plankton-Toxin Model with Inhibition of the Phytoplankton

Motivated by the discussion in Section 1, in this section we will consider the situation when toxic substance has a negative effect on the phytoplankton.

Specifically, the uptake rate and consequently the growth rate of phytoplankton is inhibited by the presence of the toxin, but zooplankton can consume the substance without any effect. An example from the field for this scenario would be a marine planktonic community comprising mainly diatoms and herbivorous copepods in a low silicate, elevated copper environment [22]. In this case copper will harm only diatoms and not crustaceans. At low concentrations the herbicide triazine also affects primary producers directly by inhibiting photosynthesis, while effects on subsequent trophic levels only would be indirect [22]. Our goal is to study toxic effects on the nutrient-plankton system by investigating simple solutions and asymptotic dynamics analytically whenever it is possible.

Let $S(t)$ denote the toxic concentration at time t. In addition to the nutrient concentration, it is assumed that the toxin is continuously pouring into the system with constant input concentration S^0 and the same constant input rate D as the nutrient. It is assumed that zooplankton can uptake the substance without any effect while phytoplankton's uptake rate of nutrient is decreased by a function $h(S)$ depending on the toxin level S. Zooplankton's toxin uptake rate is denoted by u. Functions h and u are assumed to satisfy the following assumptions.

$$\text{(H2)} h \in C^1[0,\infty), h(0) = 1, \quad h'(x) < 0 \text{ and } h(x) > 0 \quad \text{for all } x \geq 0.$$

$$\text{(H3)} u \in C^1[0,\infty), \quad u(0) = 0, \quad u'(x) > 0 \text{ for } x \geq 0 \text{ and } \lim_{x\to\infty} u(x) = 1.$$

Let $b > 0$ denote the maximum zooplankton toxin uptake rate. Similar to the previous model we assume the ecosystem under study is closed. With the above biological assumptions, the plankton-toxin interaction is given below.

$$\begin{aligned}
\dot{N} &= D(N^0(t) - N) - mf(N)h(S)P + \delta P + \epsilon Z \\
&\quad + (1-\beta)cg(P)Z + (1-\alpha)mf(N)h(S)P \\
\dot{P} &= [\alpha mf(N)h(S) - \delta - D]P - cg(P)Z \\
\dot{Z} &= [\beta cg(P) - \epsilon - D]Z \\
\dot{S} &= D(S^0 - S) - bu(S)Z \\
&\quad N(0), P(0), Z(0), S(0) \geq 0,
\end{aligned} \tag{3.1}$$

where $0 < \alpha, \beta \leq 1$ and $D, N^0, S^0, m, b, c, \epsilon, \delta > 0$.

Since $\dot{S} \leq D(S^0 - S)$ for $t \geq 0$, $\limsup_{t\to\infty} S(t) \leq S^0$. Consequently, using the same argument as we did for system (2.1), it can be easily seen that solutions of (3.1) remain nonnegative and are bounded. Moreover, system (3.1) has the following limiting system

$$\begin{aligned}
\dot{P} &= [\alpha m f(N^*(t) - P - Z)h(S) - \delta - D]P - cg(P)Z \\
\dot{Z} &= [\beta c g(P) - \epsilon - D]Z \\
\dot{S} &= D(S^0 - S) - bu(S)Z \\
& P(0), Z(0), S(0) \geq 0, P(0) + Z(0) \leq N^*(0).
\end{aligned} \tag{3.2}$$

Notice that system (3.2) is well defined as $P(t) + Z(t) \leq N^*(t)$ for $t \geq 0$ for all solutions of (3.2) with $P(0) + Z(0) \leq N^*(0)$. Clearly there always exists a trivial solution $(0, 0, S^0)$ for (3.2). Let

$$\sigma_0 = \frac{1}{\tau} \int_0^\tau [\alpha m f(N^*(t)) - \delta - D] dt$$

and

$$\rho_0 = \frac{1}{\tau} \int_0^\tau [\alpha m f(N^*(t)) h(S^0) - \delta - D] dt.$$

Then

$$\rho_0 < \sigma_0.$$

It is straightforward to show that $(0, 0, S^0)$ is locally stable if $\rho_0 < 0$. Similar to Section 2, we can show that solutions of system (3.2) asymptotically approach $(0, 0, S^0)$ if $\sigma_0 < 0$, a stronger condition than $\rho_0 < 0$.

Proposition 3.1. *If* $\sigma_0 < 0$, *then solutions of (3.2) satisfy* $\lim_{t\to\infty} P(t) = \lim_{t\to\infty} Z(t) = 0$ *and* $\lim_{t\to\infty} S(t) = S^0$.

For the autonomous case [12], numerical simulations demonstrated the existence of an attracting interior steady state when $\alpha m f(N^0) h(S^0) < \delta + D$ and $\alpha m f(N^0) > \delta + D$. Therefore, it is strongly suspected that complicated dynamical behavior can occur for system (3.2) when $\sigma_0 > 0$ and $\rho_0 < 0$. We next use numerical examples to demonstrate complexity of the model.

Let $N^0(t) = 10 + 5\sin(\frac{\pi t}{10})$, $f(x) = \frac{x}{2+x}$, $g(x) = \frac{x}{1+x}$, $h(s) = e^{-bs}$ and $u(s) = \frac{s}{6+s}$. Parameters used are $D = 0.07$, $\delta = 0.04$, $\epsilon = 0.01$, $c = 0.3$, $m = 5$, $\alpha = 0.9$, $\beta = 0.4$, $b = 1.5$ and $S^0 = 4$. In this case $\sigma_0 = 3.6369$ and $\rho_0 = -0.1007$. Therefore according to our earlier analysis that trivial solution $(0, 0, S^0)$ is locally stable. Simulations showing the existence of a positive periodic solution which is locally stable. Figure 1 provides two solutions that converge to a positive periodic solution. Figure 2 plots the trivial periodic solution $(0, 0, S^0)$ and a solution (dashed line) with initial condition $(1, 1, 5)$ that converges to the trivial periodic solution. Therefore

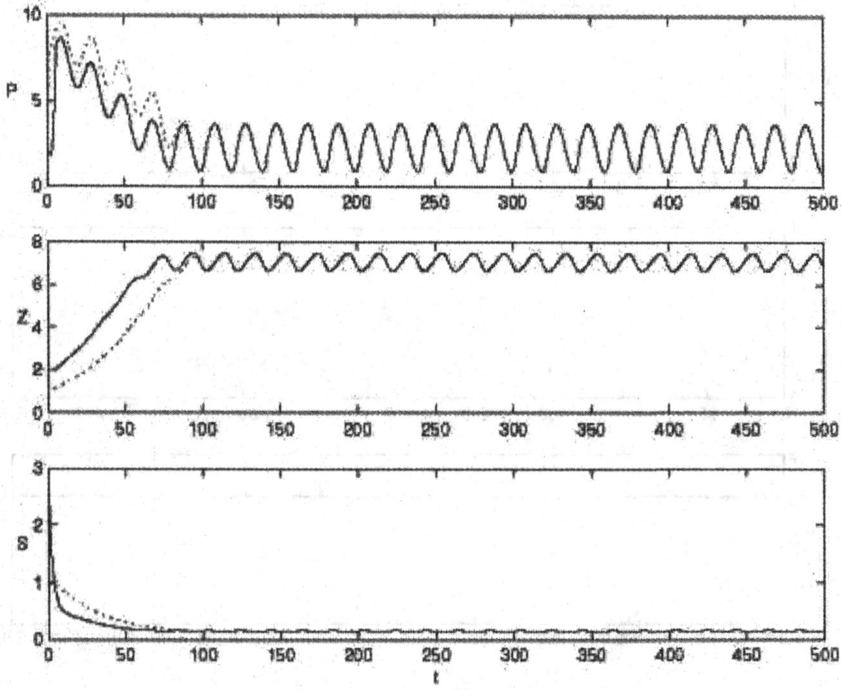

Figure 1. Both solutions asymptotically converge to the positive periodic solution. Initial conditions used are (2.5, 2, 3) for solid curve and (4, 1, 1) for dashed curve.

when $\sigma_0 > 0$ and $\rho_0 < 0$ the model exhibits a locally stable positive periodic solution even when the trivial solution is locally stable.

Proposition 3.2. *If $\rho_0 > 0$, then (3.2) has a unique τ-periodic solution of the form $(\hat{P}(t), 0, S^0)$, where $\hat{P}(t) > 0$, and solutions of (3.2) with $Z(0) = 0$ satisfy $\lim_{t\to\infty}(P(t) - \hat{P}(t)) = \lim_{t\to\infty} Z(t) = 0$ and $\lim_{t\to\infty} S(t) = S^0$.*

Proof. Since $Z(t) = 0$ for $t > 0$ if $Z(0) = 0$, it is enough to consider the following system

$$\begin{aligned} \dot{P} &= [\alpha m f(N^*(t) - P)h(S) - \delta - D]P \\ \dot{S} &= D(S^0 - S) \\ &P(0), S(0) \geq 0, P(0) \leq N^*(0). \end{aligned} \tag{3.3}$$

As $\dot{S}$ can be decoupled from P, we see that $\lim_{t\to\infty} S(t) = S^0$. Hence for any $\eta > 0$ there exists $t_0 > 0$ such that $S^0 - \eta < S(t) < S^0 + \eta$ for $t \geq t_0$. It is clear that $\frac{1}{\tau}\int_0^\tau [\alpha m f(N^*(t))h(S^0 - \eta) - \delta - D]dt > 0$. We choose $\eta > 0$

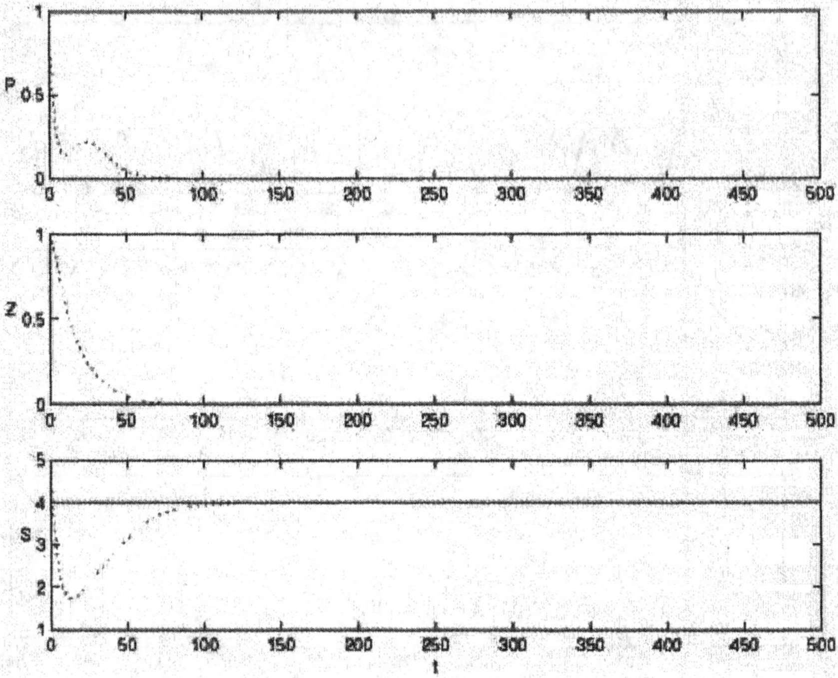

Figure 2. The figure plots the trivial solution $(0, 0, S^0)$. Another solution (dashed curve) using initial condition $(1, 1, 5)$ converges to the trivial solution.

such that

$$\frac{1}{\tau}\int_0^\tau [\alpha m f(N^*(t))h(S^0+\eta)-\delta-D]dt > 0.$$

Notice

$$\begin{aligned}\alpha m f(N^*(t)-P)h(S^0+\eta)-\delta-D]P \\ \leq \dot{P} \leq [\alpha m f(N^*(t)-P)h(S^0-\eta)-\delta-D]P\end{aligned}$$

for all $t \geq t_0$.

Considering

$$\begin{aligned}\dot{x} &= [\alpha m f(N^*(t)-x)h(S^0-\eta)-\delta-D]x \\ x(0) &= P(t_0) \leq N^*(0),\end{aligned} \tag{3.4}$$

and

$$\begin{aligned}\dot{y} &= [\alpha m f(N^*(t)-y)h(S^0+\eta)-\delta-D]y \\ y(0) &= P(t_0) \leq N^*(0).\end{aligned} \tag{3.5}$$

Let T_1 and T_2 be the Poincaré maps induced by equations (3.4) and (3.5) respectively, i.e., $T_1 : [0, N^*(0)] \to [0, N^*(0)]$ by $T_1(x_0) = x(\tau, x_0)$, where $x(t, x_0)$ is the solution of (3.4) with initial condition x_0, and T_2 is defined similarly. It follows that $T_i(0) = 0$, $\dot{T}_i > 0$, $T_i(N^*(0)) < N^*(0)$, and $\dot{T}_i(0) > 1$ for $i = 1, 2$. Thus the map T_i has a unique positive fixed point $\hat{p}^i_\eta$, $\hat{p}^i_\eta < N^*(0)$ and solutions with positive initial conditions under forward iterations of T_i all converge to $\hat{p}^i_\eta$ for $i = 1, 2$. Consequently, solutions of (3.4) and (3.5) converge to $\hat{P}^i_\eta$, where $\hat{P}^i_\eta(t)$ is the corresponding positive τ-periodic solution of (3.4) and (3.5), respectively. On the other hand $\hat{p}^i_\eta \to \hat{p}$ as $\eta \to 0^+$ for $i = 1, 2$, where $\hat{p}$ is the unique positive fixed point for the Poincaré map induced by the equation

$$\begin{aligned} \dot{P} &= [\alpha m f(N^*(t) - P)h(S^0) - \delta - D]P \\ 0 &\le P(0) \le N^*(0). \end{aligned} \tag{3.6}$$

Notice system (3.2) has a unique τ-periodic solution $(\hat{P}(t), 0, S^0)$. Since $y(t) \le P(t) \le x(t)$ for all $t \ge t_0$, we see that $\lim_{t\to\infty}(P(t) - \hat{P}(t)) = 0$ and the proof is complete. □

Let $\rho_0 > 0$ so that (3.2) has the τ-periodic solution $(\hat{P}(t), 0, S^0)$. Define

$$\rho_1 = \frac{1}{\tau}\int_0^\tau [\beta c g(\hat{P}(t)) - \delta - D]dt.$$

It is clear that $(\hat{P}(t), 0, S^0)$ is locally stable if $\rho_1 < 0$. Similar to the analysis in section 2, we are unable to reach the conclusion as whether (3.2) has a positive τ-periodic solution when $\rho_1 > 0$. We next numerically simulate the model. We adopt the same functionals as we did for the previous two plots but with somewhat different parameter values: $\alpha = 0.15$, $\beta = 0.35$ and $S^0 = 1$. In this case $\rho_0 = 0.5145 > 0$ and $\rho_1 = -0.0199 < 0$. The system has a periodic solution $(\hat{P}, 0, S^0)$ which is locally stable. Figure 3 plots three solutions with quite different behavior. The top curve using initial condition $(1, 0, 1)$ converges to the periodic solution $(\hat{P}, 0, S^0)$. The other two solutions using initial conditions $(2.5, 2, 3)$ and $(4, 1, 1)$, respectively. Therefore the system has a complicated dynamical behavior.

On the other hand, if $\rho_0 > 0$ and $\rho_1 > 0$, then apply a similar argument as in Theorem 2.4 we can show that system (3.1) is uniformly persistent.

Theorem 3.2. *If $\rho_1 > 0$ and $\rho_2 > 0$, then system (3.1) is uniformly persistent.*

We next use the same functionals as for previous graphs and choose the following parameter values: $D = 0.02$, $\delta = 0.04$, $\epsilon = 0.01$, $c = 0.3$, $m = 5$,

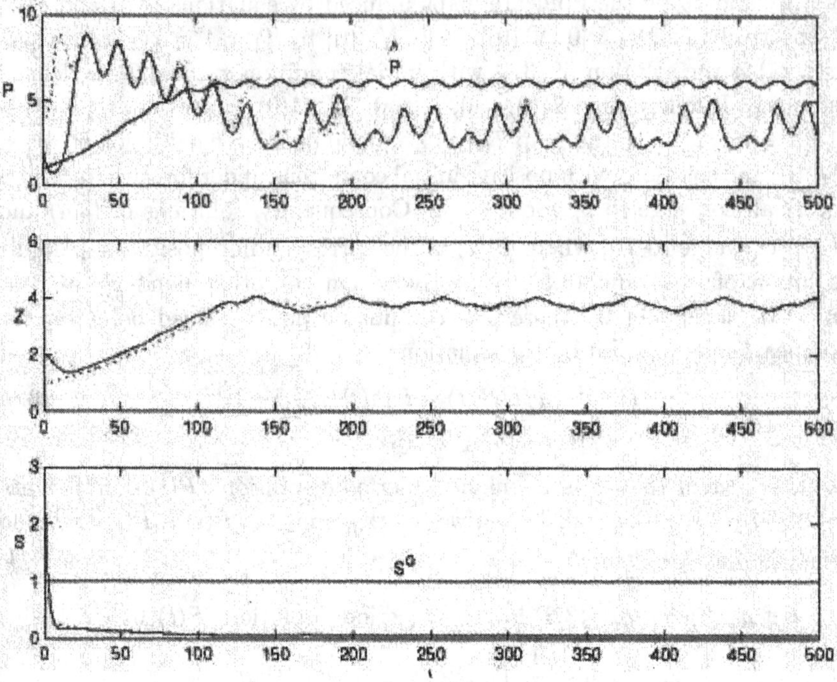

Figure 3. The figure plots three solutions. One solution with initial condition $(1, 0, 1)$ which converges to the periodic solution $(\hat{P}, 0, S^0)$. Another solution (solid curve) using initial condition $(2.5, 2, 3)$ and the other solution (dotted curve) using initial condition $(4, 1, 1)$.

$\alpha = 0.15$, $\beta = 0.35$, $b = 1.5$ and $S^0 = 1$. Then $\sigma_0 = 0.5650$, $\rho_0 = 0.0794$ and $\rho_1 = 0.0344$ and system (3.1) is uniformly persistent according to Theorem 3.2. The following figure provides two plots with initial condition $(2.5, 2, 3)$ for solid curve and $(1, 0.1, 1)$ for dashed curve. Although solutions are oscillating, both plankton populations survived.

4. Discussion

Nutrient-phytoplankton-zooplankton models are proposed to study the effects of pollutants upon the nutrient-plankton interaction. For simplicity, the nutrient-plankton interaction is assumed to be a closed ecological system. The input nutrient concentration motivated by the seasonal and day/night cycles is assumed to be input periodically. However, the toxin is continuously pouring into the system with a constant concentration. It was shown analytically that there exist population thresholds for the model

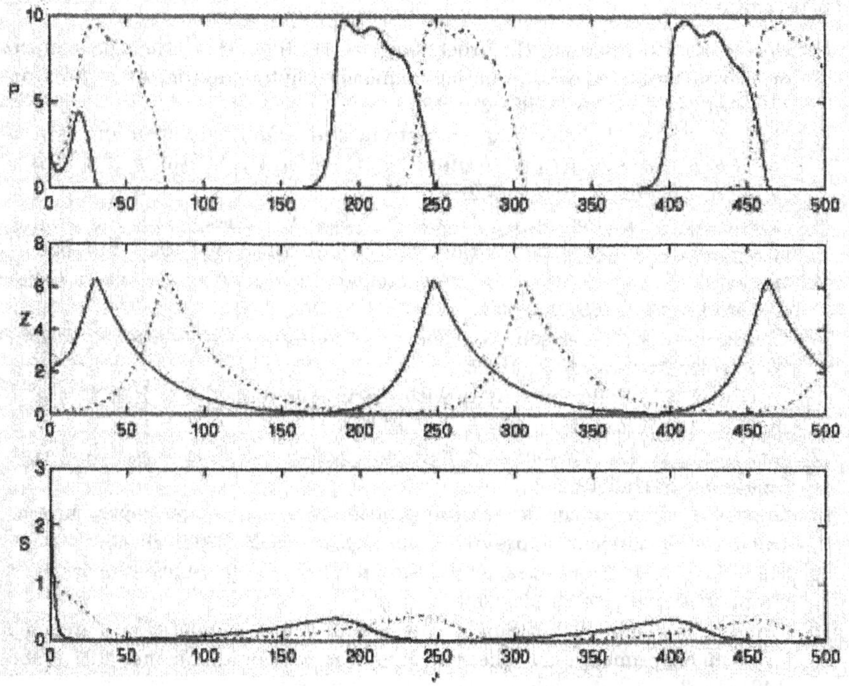

Figure 4. The figure plots two solutions. One solution (solid curve) with initial condition $(2.5, 2, 3)$ and the other solution (dashed curve) using initial condition $(1, 0.1, 1)$.

without the toxin. Both populations can coexist if the lumped parameters σ_0 and σ_a are positive. When $\sigma_0 < 0$, then both populations go to extinction. Only the phytoplankton population can survive if $\sigma_0 > 0$ and $\sigma_1 < 0$.

The introduction of an inhibited substance can alter the dynamical behavior of the plankton interaction unpredictably. The survival and/or extinction of the populations are initial condition dependent. Unlike the model without the toxin, phytoplankton may survive even if $\rho_0 < 0$. This is very counter-intuitive as the growth rate of the phytoplankton is diminished due to the toxic substance. Therefore it needs a more delicate ecological study to understand the interaction of plankton populations when pollution is present, especially in the area when phytoplankton population is small but with large concentration of nutrient. It would be interesting to compare the minimum and maximum values of these periodic solutions with the model of no toxin. What happens when the inhibition occurring in the upper trophic level is also worth of pursuing. We shall leave these questions as another research project to study.

References

[1] Bester, K., Huehnerfuss, H., Brockmann, U. H., Rick, H. J., Biological effects of triazine herbicide contamination on marine phytoplankton. *Arch. Environ. Contam. Toxicol*, **29** (1995) 277–283.

[2] Bull, A., Slater, J., Microbial interactions and community structure, *Microbial Interactions and Communities*, Vol.1, edited by A. Bull and J. Slater, London: Academic Press, 1982.

[3] Busenberg, S., Kumar, S. K., Austin, P., Wake, G., The dynamics of a model of a plankton-nutrient interaction. *Bull. Math. Biol.*, **52** (1990) 677–696.

[4] Butler, G. J. and Waltman, P., Persistence in dynamical systems. *J. Differ. Equations*, **63** (1986) 255–262.

[5] Coddington, E., Levinson, N., *Theory of Ordinary Differential Equations*, New York: McGraw Hill, 1955.

[6] Davies, A. G., Pollution studies with marine phytoplankton. Part II. Heavy metals. *Adv. Mar. Biol.*, **15** (1978) 381–508.

[7] DeAngelis, D. L., *Dynamics of Nutrient Cycling and Food Webs.* New York: Chapman & Hall 1992.

[8] Grover, J., The impact of variable stoichiometry on predator-prey interactions: a multinutrient approach. *Amer. Natur.*, **162** (2003) 29–43.

[9] Hale, J. K. and Somolinos, A. S., Competition for fluctuating nutrient. *J. Math. Biol.*, **18** (1983) 255–280

[10] Hansen, S., Hubbell, S., Single nutrient microbial competition: agreement between experimental and theoretical forecast outcomes, *Science*, **207** (1980) 1491–1493.

[11] Jang, S. R.-J., Baglama, J., Persistence in variable-yield nutrient-plankton models with nutrient recycling. *Mathematical and Computer Modelling*, **38** (2003) 281–298.

[12] Jang, S. R.-J., Baglama, J., Rick, J., Nutrient-phytoplankton-zooplankton models with a toxin, *Math. Comput. Mod.*, to appear.

[13] Kreutzweiser, D., Back, R., Sutton, T., Thompson, D., Scarr, T., Community-level disruptions among zooplankton of pond mesocosms treated with a neem (azadirachtin) insecticide, *Aquatic Toxicology*, **56** (1998) 257–273.

[14] Lalli, C. M., Parsons, T. R., *Biological Oceanography: An introduction*, Butterworth-Heinemann, 1997.

[15] Lenski, R., Hattingh, S., Coexistence of two competitors on one resource and one inhibitor: A chemostat model based on bacteria and antibiotics, *J. Theor. Biol.*, **122** (1986) 83–93.

[16] Rand, G., Clark, J., Holmes, C., Use of outdoor freshwater pond microcosmos: II. responses of biota to pyridan, *Environmental Toxicology & Chemistry*, **19** (2000) 398–404.

[17] Rick, H. J., Repercussions of the silicate copper interaction in marine diatoms on planktonic systems. Habilitation Thesis, University of Kiel, Germany, 2000.

[18] Rick, H. J., Rick, S., Anthropogenic distorted Si-Cu ratios — effects on coastal plankton communities. Presentation at SETAC 23rd Annual

Meeting: Achieving Global Environmental Quality: Integrating Sciene & Management, 16–20 Nov. 2002, Salt Lake City, Utah, 2002.

[19] Riedel, G. F., Influence of salinity and sulfate on the toxicity of Cr(VI) to the estuarine diatom Thalassiosira pseudonana. J. Phycol., **20** (1998) 496–500.

[20] Ruan, S., Persistence and coexistence in zooplankton-phytoplankton-nutrient models with instantaneous nutrient recycling. *J. Math. Biol.*, **31** (1993) 633–654.

[21] Ruan, S., Oscillations in plankton models with nutrient recycling. *J. Theor. Biol.*, **208** (2001) 15–26.

[22] Rueter, J. R., Chisholm, S. W., Morel, F. M. M., Effects of copper toxicity on silicon acid uptake and growth in Thalassiosira pseudonana. *J. Phycol.*, **17** (1981) 270–278.

[23] Sunda, W. G., Huntsman, S. A., Processes regulating cellular metal accumulation and physiological effects. Phytoplankton as model systems. *Sci. Total Environ.*, **219** (1998) 165–181.

[24] Sunda, W. G., Huntsman, S. A., Interactive effects of external manganese, the toxic metals copper and zinc, and light in controlling cellular manganese and growth in a coastal diatom. *Limnol. Oceanogr.*, **43** (1998) 1467–1475.

[25] Thomas, W. H., Seibert, D., Effects of copper on the dominance and the diversity of algae: Controlled ecosystem pollution experiment. *Bull. Mar. Sci*, **27**(1) (1977) 23–33.

[26] Werner, I., Hinton, D., Bailey, H., Connor, V., De Vlaming, V., Deanovic, L., Insecticide-caused toxicity to Ceriodaphnia dubia (Cladocera) in the Sacramento-San Joaquin River Delta, California, *Environmental Toxicology & Chemistry*, **19** (2000) 215.

Monitoring [illegible] Global Environment [illegible] *In [illegible]* [illegible] 16 [illegible] 2004, [illegible] (2002) [illegible]

[14] [illegible] Influence of [illegible] and [illegible] on the [illegible] of the [illegible] Thai [illegible] (1992) [illegible]

[15] [illegible] and [illegible] nutrient [illegible] *J.* [illegible] 31 [illegible] 653–664.

[16] [illegible] S. [illegible] with nutrient [illegible] 308 (2004) [illegible]

[17] [illegible] R. [illegible] W. [illegible] F. [illegible] M. [illegible] of [illegible] on [illegible] in [illegible] *J.* [illegible] 17 [illegible] 278.

[18] [illegible] W. [illegible] S.A. [illegible] of [illegible] 219 [illegible]

[19] [illegible] W. [illegible] S.A. [illegible] and [illegible] 41 [illegible]

[20] [illegible] D. [illegible] Effects [illegible]

[21] [illegible] (2001) [illegible] *Proc.* [illegible] V. [illegible] Delta [illegible] (2003) [illegible]

ON THE MECHANISM OF STRAIN REPLACEMENT IN EPIDEMIC MODELS WITH VACCINATION

MAIA MARTCHEVA

Department of Mathematics,
University of Florida, 358 Little Hall,
PO Box 118105, Gainesville, FL 32611-8105

August 13, 2006

Strain replacement is the effect of substitution of a strain of higher prevalence in the population with another. Differential effectiveness of the vaccines is thought to be the mechanism responsible for the replacement effect. Recent theoretical study shows that differential effectiveness of the vaccine may not be necessary and other trade-off mechanisms can lead to it even when the vaccine is "perfect". We suggest that the mechanism of strain replacement is the reciprocal effect of vaccination on the fitness of the strains as measured by their invasion reproduction numbers. This mechanism is responsible for the substitution of one strain with another to occur both when the vaccine is perfect and when it is imperfect. We review various well-known trade-off mechanisms and investigate whether they lead to replacement effect in conjunction with "perfect" vaccination. We find that in contrast to imperfect vaccination which leads to replacement of a strain with larger intrinsic reproduction number with a strain with a lower intrinsic reproduction number, "perfect" vaccination seems to have opposite effect on the intrinsic reproduction numbers.

Keywords: multiple pathogen variants, strain replacement, coinfection, cross-immunity, vaccination, coexistence, invasion.

1. Introduction

In response to selective pressures from the host immune system pathogens vary their genetic characteristics to escape recognition. Thus the evolution and replacement of pathogen types is a continuous process mediated by the host immunity. The rate at which a pathogen mutant obtains dominance in the individual host is highest at intermediate level of immunity of the host Vaccination has direct impact on host immunity and is therefore intimately connected to the evolution of pathogens on the within-host level. Furthermore, vaccination changes dynamically the susceptible pool for the pathogen variants on population level and is a mechanism that favors the population distribution of a certain strain. The process through which the establishment of a particular pathogen variant on within host level is related to the establishment of this or other pathogen variant on the population level but this relation is not well understood. Phylogenies of specific highly mutable pathogens (such as HIV) show significant differences in the evolution on the within-host and between-host levels [12].

Vaccination plays distinctive role in the evolution of the pathogens on each level [23], but its role as an evolutionary agent is better understood on population level. Empirical evidence in terms of clinical trials and surveillance [4, 15] as well as theoretical research [8, 19, 20, 27, 28] point to the fact that while vaccination leads to elimination of certain strains also facilitates the emergence at higher prevalence of strains which previously were not widely spread. This phenomenon is now called *the replacement effect* [24]. The replacement effect has been drawing significant attention in the literature because it diminishes the effect of vaccination, particularly for diseases caused by pathogens of considerable genetic diversity. Clearly, vaccines should be developed in a way that minimizes the possibility for substitution of the current strains with others. This, in turn, requires that we understand what causes this effect.

The primary reason for strain replacement to occur is that vaccines do not equally well protect against all strains — a property referred to as *differential effectiveness.* In a recent article we [16] showed, however, that differential effectiveness may not be necessary for the replacement to occur. This raises the question of the mechanism of strain replacement — a mechanism that can explain its occurrence both in the presence and the absence of differential effectiveness. In this article, namely in the next section, we discuss such a mechanism strictly in the case of *strong replacement effect* — that is, replacement effect in which the dominance of one strain is exchanged with dominance of the other.

Further we observe that what is necessary for replacement to occur is the action of some sort of *trade-off mechanism* — a mechanism that allows for coexistence. In section 3 we show that differentially effective vaccines themselves are a trade-off mechanism while equally effective, and in particular "perfect" vaccines, that provide complete protection, lead to competitive exclusion. That observation explains why equally effective vaccines when acting outside of other trade-off mechanisms cannot cause replacement. However, the results in [16] show that equally effective vaccines can act synergistically with some other trade-off mechanism (super-infection in that case) to lead to strain replacement. Can all trade-off mechanisms fill that role or there is something special about super-infection? We devote the rest of the paper to answering this question. In section 4 we investigate coinfection coupled with "perfect" vaccination and we find that strain replacement can also occur. In section 5 we investigate cross-immunity and we touch on mutation. We find that with these two trade-off mechanisms strain replacement in its strong form does not occur. Our results are in accord with those in [13] where cross-immunity has been found to lead to selection for a subdominant strain only in presence of imperfect vaccine.

Strain replacement signifies pathogen evolution under the influence of vaccination. One of the questions that arise in that context concerns the direction of this evolution. Is it possible that through vaccination we may be selecting for a more virulent strain? We do not address this question here but it has been investigated for incompletely effective vaccines in [11] where it has been observed that virulence can evolve towards increased virulence or decreased virulence depending on whether the vaccine blocks pathogen growth or infection. Here we observe that the direction of evolution of the pathogen's intrinsic reproduction number (the reproduction number in the absence of vaccination) depends on whether the vaccine is differentially effective or not. Imperfect vaccines seem to lead to evolution towards lower intrinsic reproduction numbers while "perfect" vaccines lead to evolution to higher reproduction numbers. In section 6 we summarize our results and discuss the differences in the the replacement effect for perfect and imperfect vaccines.

2. The Mechanism of Strain Replacement

Intuitively, replacement on population level of one pathogen strain with another suggests exchange of prevalence between the two strains. Such exchange of prevalence is a result of vaccination and can occur under several scenarios which are mathematically distinct.

- **Scenario 1:** The strains coexist both before and after vaccination but before vaccination strain one is more prevalent while after vaccination strain two is more prevalent. We will refer to this replacement as *weak replacement.*
- **Scenario 2:** Strain one dominates (persists alone) before vaccination but after vaccination the two strains coexist but strain two has higher prevalence. An analogous scenario occurs if both strains coexist before vaccination with strain one being more prevalent, while after vaccination strain two dominates.

- **Scenario 3:** Strain one dominates (persists alone) before vaccination while after vaccination strain two dominates. We will call this strain replacement *strong replacement.*

Replacement in all those three scenarios may or may not be epidemiologically significant. Replacement will be epidemiologically significant if the prevalence of the second strain after vaccination is about or higher than the prevalence of the first strain before vaccination, or, in other words, the prevalence of the replacing strain is sufficiently high.

Mathematically scenario 3 is easier to investigate than scenarios one and two because we have strict rigorous conditions which allow the prediction of which strain will dominate. On the other hand to investigate the other two scenarios we need to know which strain will have higher prevalence if the two coexist — something that is not that well understood.

The conditions that govern the dominance of one strain or another are based on the invasion reproduction numbers — the number of cases strain i will generate when strain j is at equilibrium. We denote the invasion reproduction number of strain i by $\hat{\mathcal{R}}_i$. Thus, strain one dominates if strain one can invade the equilibrium of strain two, $\hat{\mathcal{R}}_1 > 1$ while strain two cannot invade the equilibrium of strain one, $\hat{\mathcal{R}}_2 < 1$. Analogous condition determine the dominance of strain two. The two invasion reproduction numbers depend on the vaccination level ψ: $\hat{\mathcal{R}}_i(\psi)$. Suppose that strain one dominates in the absence of vaccination, that is, $\hat{\mathcal{R}}_1(0) > 1$ while $\hat{\mathcal{R}}_2(0) < 1$. In order for strain two to dominate at a certain vaccination level $\hat{\psi}$ we need that strain one cannot invade the equilibrium of strain two, $\hat{\mathcal{R}}_1(\hat{\psi}) < 1$, and that strain two can invade the equilibrium of strain one, $\hat{\mathcal{R}}_2(\hat{\psi}) > 1$. If such a vaccination level $\hat{\psi}$ exists, strong replacement will occur. In other words we need vaccination to reduce the invasion capabilities of the first strain and to increase invasion capabilities. Without such reciprocal effect of vaccination, strong replacement cannot occur.

Fitness is defined as the expected number of offspring contributed to the next generation and is typically computed as the product of survival and average number of offspring. For that reason fitness of pathogens is often identified as their reproduction number. However, the reproduction numbers are a measure of the reproductive success of the pathogen in an entirely susceptible population, and, in a presence of a competitor, higher reproduction number may not even lead to persistence [21, 25].

The invasion reproduction numbers, on the other hand, measure reproductive success when the competitor is established, and guarantee persistence (at the very least). Thus, they reflect better the fitness of the pathogens. If we take the invasion reproduction numbers for a measure of the fitness, then the mechanism for strain replacement says that vaccination must have a reciprocal (differential in direction) effect on the fitness of the pathogens for strong replacement to occur. Strong replacement cannot occur without such differential in direction effect on the fitness of the pathogens.

3. Differential Effectiveness of the Vaccine and Strain Replacement

In this section we show two things. First, that equally effective vaccines cannot cause coexistence, that is, if competitive exclusion is the norm in the absence of vaccination, it is also the only outcome in the presence of vaccination with equally effective vaccine. In contrast, differentially effective vaccines can cause coexistence, even if in the absence of vaccination competitive exclusion is the only outcome, that is differential effectiveness of the vaccine is a trade-off mechanism in its own right. Second, we show that differential effectiveness of the vaccine alone leads to reciprocal impact of vaccination on the invasion reproduction numbers of the pathogens and therefore, to strain replacement. Consequently, differential effectiveness of the vaccine is one manifestation of the main mechanism causing strain replacement.

We consider a host population of total size at time t given by $N(t)$ that is being recruited at a rate Λ and dies at a natural death rate μ. The number of healthy individuals who are susceptible to the disease at time t is denoted by $S(t)$. Healthy individuals can get infected by strain one at a transmission rate β_1 and enter the class of individuals infected and infectious with strain one. This class is of total size $I(t)$. Independently, healthy individuals can get infected by strain two at a transmission rate β_2 and enter the class of individuals infected and infectious with strain two whose total size is given by $J(t)$. Infected individuals with strain one recover at a recovery rate γ_1 while infected individuals with strain two recover at a recovery rate γ_2. Recovered individuals comprise the class $R(t)$. Finally, susceptible individuals are vaccinated at a vaccination rate ψ and enter the class of vaccinated individuals, $V(t)$. We assume that vaccinated individuals can get infected by strain one at a rate $\beta_1\delta_1$ where δ_1 is the coefficient of reduction of transmission of strain one provided by the vaccine. Similarly, vaccinated individuals can get infected by strain two at a rate $\beta_2\delta_2$ where δ_2 is the coefficient of reduction of transmission of strain two provided by the vaccine. We will consider two cases mainly.

(1) The vaccine is equally effective with respect to both strains, that is, $\delta_1 = \delta_2 = \delta$ which may or may not be zero.
(2) The vaccine has differential effectiveness. In particular, we will assume that the vaccine is nearly perfect with respect to one of the strains, say strain two. That means that $\delta_2 = 0$. In contrast, the vaccine is only partially effective with respect to strain one, that is, $\delta_1 \neq 0$. We denote $\delta_1 = \delta$.

We consider the following two-strain model with vaccination [14].

$$\begin{aligned}
S' &= \Lambda - \beta_1 \frac{SI}{N} - \beta_2 \frac{SJ}{N} - (\mu + \psi)S \\
I' &= \beta_1 \frac{SI}{N} + \beta_1 \delta_1 \frac{IV}{N} - (\mu + \gamma_1)I \\
J' &= \beta_2 \frac{SJ}{N} + \beta_2 \delta_2 \frac{JV}{N} - (\mu + \gamma_2)J \\
R' &= \gamma_1 I + \gamma_2 J - \mu R \\
V' &= \psi S - \beta_1 \delta_1 \frac{IV}{N} - \beta_2 \delta_2 \frac{JV}{N} - \mu V
\end{aligned} \tag{3.1}$$

Since vaccines are generally assumed to reduce transmission we must have $0 \leq \delta_1, \delta_2 \leq 1$. The reproduction numbers of the two strains are given by

$$\mathcal{R}_1(\psi) = \frac{\beta_1 \mu + \beta_1 \delta_1 \psi}{(\mu + \psi)(\mu + \gamma_1)} \qquad \mathcal{R}_2(\psi) = \frac{\beta_2 \mu + \beta_2 \delta_2 \psi}{(\mu + \psi)(\mu + \gamma_2)} \tag{3.2}$$

We note that both reproduction numbers are decreasing functions of the vaccination rate ψ. We denote the value of the reproduction numbers in the absence of vaccination, $\mathcal{R}_i(0)$ at $\psi = 0$, with $\mathcal{R}_i$ and we call those *intrinsic reproductive numbers.* Furthermore, the value at maximal vaccination levels, $\psi \to \infty$, is $\mathcal{R}_i(0)\delta_i$, and may or may not be under one. This reflects the fact that imperfect vaccines may not be able to reduce the reproduction number below one, and may not lead to eradication. The system above always has a disease-free equilibrium (each equilibrium is given in terms of the proportions of susceptible, infectives with each strain, recovered and vaccinated individuals – $\mathcal{E} = (s, i, j, r, v)$):

$$\mathcal{E}_0 = \left(\frac{\mu}{\mu + \psi}, 0, 0, 0, \frac{\psi}{\mu + \psi} \right),$$

and a unique dominance equilibrium corresponding to each strain. The dominance equilibrium of the first strain is

$$\mathcal{E}_1 = \left(\frac{\mu}{\beta_1 i + \mu + \psi}, i, 0, \frac{\gamma_1 i}{\mu}, \frac{\psi \mu}{(\beta_1 \delta_1 i + \mu)(\beta_1 i + \mu + \psi)} \right)$$

and it exists when $\mathcal{R}_1(\psi) > 1$. The proportion of infected with strain one i in $\mathcal{E}_1$ is given by the unique solution of the following equation:

$$\frac{\mathcal{R}_1\mu}{\beta_1 i + \mu + \psi}\left[1 + \frac{\delta_1\psi}{\beta_1\delta_1 i + \mu}\right] = 1. \tag{3.3}$$

The dominance equilibrium of the second strain, correspondingly, is

$$\mathcal{E}_2 = \left(\frac{\mu}{\beta_2 j + \mu + \psi}, 0, j, \frac{\gamma_2 j}{\mu}, \frac{\psi\mu}{(\beta_2\delta_2 j + \mu)(\beta_2 j + \mu + \psi)}\right)$$

and it exists when $\mathcal{R}_2(\psi) > 1$. The system (3.1) may or may not have coexistence equilibria.

Our first result testifies to the fact that competitive exclusion is the only outcome in the absence of vaccination.

Proposition 3.1. *Assume $\psi = 0$. Then, if $\max\{\mathcal{R}_1, \mathcal{R}_2\} > 1$, a competitive exclusion principle holds, that is, the strain with the larger reproduction number persists, while the other one is eliminated.*

This results follows from the observation that $\psi = 0$ implies that $V(t) \to 0$ as $t \to \infty$. The rest of the system is similar to the one studied in [3] and the result follows from there. Our next result shows that an equally effective vaccine $\delta_1 = \delta_2 = \delta$ also leads to competitive exclusion, namely,

Proposition 3.2. *Assume that the vaccine is equally effective with respect to both strains, that is, $\delta_1 = \delta_2 = \delta$. Then, if $\max\{\mathcal{R}_1(\psi), \mathcal{R}_2(\psi)\} > 1$, a competitive exclusion principle holds, that is, the strain with the larger reproduction number persists, while the other one is eliminated.*

Proof. Assume without loss of generality that $\mathcal{R}_1(\psi) > \mathcal{R}_2(\psi)$. As a result of the assumption that $\delta_1 = \delta_2 = \delta$ this inequality is equivalent to the inequality $\beta_1(\mu + \gamma_2) > \beta_2(\mu + \gamma_1)$. Consider the function $\xi(t) = I^{\beta_2}(t)/J^{\beta_1}(t)$. Differentiating ξ with respect to t we see that it satisfies the following differential equation $\xi'(t) = \alpha\xi(t)$ where the constant α is given by

$$\alpha = [\beta_1(\mu + \gamma_2) - \beta_2(\mu + \gamma_1)] = \frac{\mu + \psi}{\mu + \delta\psi}[\mathcal{R}_1(\psi) - \mathcal{R}_2(\psi)] > 0$$

Consequently, $\xi(t) \to \infty$ as $t \to \infty$, and since $I(t)$ is bounded, we must have $J(t) \to 0$ and $t \to \infty$. That implies persistence of $I(t)$ (at least in a weak sense) because if we assume $I(t) \to 0$, then the solutions of the system (3.1) approach the disease-free equilibrium, which on the other hand can be shown to be unstable because at least one of the reproduction numbers is above one. Therefore, the assumption that $I(t) \to 0$ is not correct. □

Now we turn to the scenario (2): differential effectiveness of the vaccine. We work under the conditions $\delta_2 = 0$ and $\delta_1 = \delta$. The model (3.1) takes the form:

$$\begin{aligned} S' &= \Lambda - \beta_1 \frac{SI}{N} - \beta_2 \frac{SJ}{N} - (\mu + \psi) S \\ I' &= \beta_1 \frac{SI}{N} + \beta_1 \delta \frac{IV}{N} - (\mu + \gamma_1) I \\ J' &= \beta_2 \frac{SJ}{N} - (\mu + \gamma_2) J \\ R' &= \gamma_1 I + \gamma_2 J - \mu R \\ V' &= \psi S - \beta_1 \delta \frac{IV}{N} - \mu V \end{aligned} \tag{3.4}$$

In this case equation (3.3) for j simplifies and gives the following solution:

$$j = \frac{(\mu + \psi)(\mathcal{R}_2(\psi) - 1)}{\beta_2}$$

and dominance equilibrium $\mathcal{E}_2$ takes the form

$$\mathcal{E}_2 = \left(\frac{1}{\mathcal{R}_2}, 0, \frac{(\mu + \psi)(\mathcal{R}_2(\psi) - 1)}{\beta_2}, \frac{\gamma_2}{\mu} \frac{(\mu + \psi)(\mathcal{R}_2(\psi) - 1)}{\beta_2}, \frac{\psi}{\mathcal{R}_2 \mu} \right).$$

To investigate the possible presence of a coexistence equilibrium, we introduce the invasion reproduction numbers. The invasion reproduction number of the first strain at the equilibrium of the second strain $\hat{\mathcal{R}}_1(\psi)$ is the number of secondary infections one individual infected with the first strain can produce when the second strain is at equilibrium in the population. This number under scenario (2) takes the form

$$\hat{\mathcal{R}}_1(\psi) = \frac{\mathcal{R}_1(\mu + \delta\psi)}{\mathcal{R}_2 \mu} = \frac{\mathcal{R}_1(\psi)}{\mathcal{R}_2(\psi)} \tag{3.5}$$

where $\mathcal{R}_i = \mathcal{R}_i(0)$ for $i = 1, 2$. Similarly, the invasion reproduction number of the second strain at the equilibrium of the first strain $\hat{\mathcal{R}}_2(\psi)$ is the number of secondary infections one individual infected with the second strain can produce when the second strain is at equilibrium in the population. The invasion reproduction number of the second strain under scenario (2) takes the form

$$\hat{\mathcal{R}}_2(\psi) = \frac{\mathcal{R}_2 \mu}{\beta_1 i + \mu + \psi} \tag{3.6}$$

where i is the solution of (3.3). First we show that under scenario (2) competitive exclusion is not necessarily the outcome, and coexistence is possible. In other words unequally effective vaccines represent a trade-off mechanism which allows for coexistence. The following result testifies to the presence of a nontrivial region of the parameter space where coexistence may occur.

Proposition 3.3. *If $\mathcal{R}_2 > \mathcal{R}_1$, and each strain can invade the equilibrium of the other, that is*

$$\hat{\mathcal{R}}_1(\psi) > 1 \qquad \hat{\mathcal{R}}_2(\psi) > 1$$

then there exists a unique coexistence equilibrium $\mathcal{E}^* = (s^*, i^*, j^*, r^*, v^*)$.

Proof. The values of the coexistence equilibrium can be computed as follows

$$\begin{aligned} s^* &= \frac{1}{\mathcal{R}_2} \\ i^* &= \frac{\mathcal{R}_2\mu(\hat{\mathcal{R}}_1(\psi)-1)}{(\mathcal{R}_2-\mathcal{R}_1)\beta_1\delta} \\ j^* &= \frac{1}{\beta_2}[\mathcal{R}_2\mu-(\beta_1 i^*+\mu+\psi)] \\ r^* &= \frac{\gamma_1 i^*}{\mu}+\frac{\gamma_2 j^*}{\mu} \\ v^* &= \frac{\mathcal{R}_2-\mathcal{R}_1}{\delta\mathcal{R}_1\mathcal{R}_2} \end{aligned} \tag{3.7}$$

which gives the uniqueness. The existence will follow if all values above are positive. This is straight forward to see under the assumptions of the this proposition for all values except j^*. To see that $j^* > 0$ consider the left-hand side of equation (3.3) as follows:

$$f(x,y) = \frac{\mathcal{R}_1\mu}{\beta_1 x+\mu+\psi}\left[1+\frac{\delta\psi}{\beta_1\delta y+\mu}\right].$$

The fact that i — the proportion of infectives with strain one in $\mathcal{E}_1$ — means that i satisfies equation (3.3), that is $f(i,i) = 1$. On the other hand if i^* is the proportion of infectives in the coexistence equilibrium $\mathcal{E}^*$, then we have

$$\beta_1\delta i^* + \mu = \frac{\delta\psi\mathcal{R}_1}{\mathcal{R}_2-\mathcal{R}_1}$$

and consequently, $f(i,i^*) = \hat{\mathcal{R}}_2(\psi) > 1$. Therefore, $f(i,i) < f(i,i^*)$. But $f(x,y)$ is a decreasing function of y which gives $i^* < i$. Then,

$$j^* = \frac{\beta_1 i^*+\mu+\psi}{\beta_2}\left[\frac{\mathcal{R}_2\mu}{\beta_1 i^*+\mu+\psi}-1\right] > \frac{\beta_1 i^*+\mu+\psi}{\beta_2}\left[\hat{\mathcal{R}}_2(\psi)-1\right] > 0.$$

The persistence of both strains in this case can be observed in simulations. □

Several remarks are in order. First, we note that $\hat{\mathcal{R}}_1(\psi) > 1$ is equivalent to $\mathcal{R}_1(\psi) > \mathcal{R}_2(\psi)$. This, in particular means that in the absence of vaccination, $\psi = 0$, the conditions of this proposition are inconsistent and coexistence does not occur. If we have $\mathcal{R}_2 > \mathcal{R}_1$ we need the vaccination level $\psi > \psi^*$, where the threshold vaccination level ψ^* is given by

$$\psi^* = \frac{(\mathcal{R}_2 - \mathcal{R}_1)\mu}{\mathcal{R}_1 \delta}$$

so that $\mathcal{R}_1(\psi) > \mathcal{R}_2(\psi)$. Second, $\hat{\mathcal{R}}_2(\psi) > 1$ implies that $\mathcal{R}_2(\psi) > 1$. Thus, the conditions of the proposition imply that both reproduction numbers are above one. Consequently subthreshold coexistence does not occur.

Next, we turn our attention to strain replacement. First, we note that vaccination has a reciprocal effect on the two invasion reproduction numbers. In particular, $\hat{\mathcal{R}}_1(\psi)$ is a linearly increasing function of the vaccination rate ψ such that $\hat{\mathcal{R}}_1(0) = \frac{\mathcal{R}_1}{\mathcal{R}_2}$. That is, under one of the assumptions for coexistence, $\mathcal{R}_2 > \mathcal{R}_1$, the invasion reproduction number of the first strain in the absence of vaccination is smaller than one. It is somewhat complicated to express the invasion reproduction number of the second strain as a function of the vaccination rate. Instead, we will use an upper and a lower bound of that number as follows:

$$\frac{\mathcal{R}_2 \mu}{\beta_1 + \mu + \psi} \leq \hat{\mathcal{R}}_2(\psi) \leq \frac{\mathcal{R}_2 \mu}{\mu + \psi} \tag{3.8}$$

From these inequalities, it can be seen that $\hat{\mathcal{R}}_2(\psi) \to 0$ as $\psi \to \infty$ although it may not do so monotonically. This, in particular, implies that vaccination has a reciprocal effect on the invasion capabilities of the two strains — it increases the invasion capabilities of the first strain, and decreases the invasion capabilities of the second strain. We illustrate in Figure 1 the graph of the invasion reproduction number of the first strain and we plot the upper bound and the lower bound from (3.8) to limit the region that contains the invasion reproduction number of the second strain.

In the case of absence of vaccination, $\psi = 0$, the proportion infectives i can be easily computed from (3.3) and it can be seen that $\hat{\mathcal{R}}_2(0) = \frac{\mathcal{R}_2}{\mathcal{R}_1}$. That is, under one of the assumptions for coexistence, $\mathcal{R}_2 > \mathcal{R}_1$, the invasion reproduction number of the second strain in the absence of vaccination

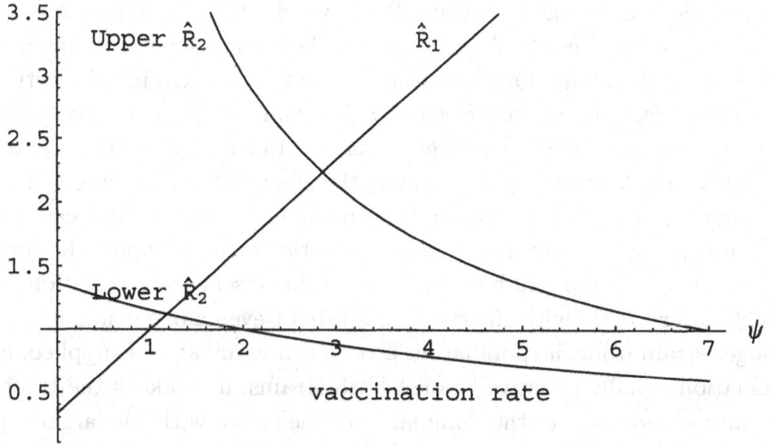

Figure 1. The figure illustrates the graph of the invasion reproduction number $\hat{\mathcal{R}}_1(\psi)$ and Upper $\hat{\mathcal{R}}_2(\psi)$ give the upper bound while Lower $\hat{\mathcal{R}}_2(\psi)$ gives the lower bound in (3.8). The parameters are chosen as follows: $\beta_1 = 5$, $\beta_2 = 15$, $\gamma_1 = 0.5$, $\gamma_2 = 0.5$, $\delta = 1$, $\mu = 0.5$ which give $\mathcal{R}_1 = 5$ and $\mathcal{R}_2 = 15$.

is greater than one. Consequently, in the absence of vaccination we have the second strain dominating in the population. As the vaccination levels increase, the invasion capabilities of the first strain grow while the invasion capabilities of the second decline until, at some vaccination level $\hat{\psi}$ we have $\hat{\mathcal{R}}_1(\hat{\psi}) > 1$ and $\hat{\mathcal{R}}_2(\hat{\psi}) < 1$, that is the first strain dominates in the population. Looking at Figure 1 we can choose $\hat{\psi}$ to be any value greater than seven. A replacement effect has occurred.

4. Coinfection, Perfect Vaccination and Strain Replacement

In an earlier article [16] we have reported that super-infection as a trade-off mechanism can lead to strain replacement, even when the vaccine is assumed perfect with respect to both strains, that is, it protects all vaccinated individuals completely from infection with either strain. In super-infection one of the strains (say, strain one) wins instantaneously the within-host competition and displaces the other strain (say, strain two) in infected individuals thus turning individuals infected with the second strain into individuals infected with the first strain [25]. If the second strain has better reproduction rate and would be the dominant strain in the absence of vaccination and super-infection, super-infection as a trade-off mechanism might be strengthening the first strain to coexist with the second strain, or even replace it as a dominant strain. When vaccination is applied against the first strain, it weakens the first strain and the second strain can again take over, that is, dominate. One question we want to address in this section is whether coinfection, as another trade-off mechanism [25] can lead to strain replacement even when the vaccine is perfect. In coinfection the strains are seemingly symmetrical — they can both infect individuals infected with the other strain and then they coexist in the host and the host can transmit both. As a trade-off mechanism coinfection can also allow the weaker strain — the one that will be eliminated in the absence of coinfection — to persist either jointly with the stronger strain or even alone, eliminating the stronger strain from the population [21]. When vaccination is applied, even vaccination equally effective against both strains, it weakens the trade-off mechanism and restores the dominance of the strain with the larger reproduction number in absence of vaccination. Thus the same effect is observed when the trade-off mechanism is coinfection and leads to strain replacement as we show below.

We consider again a host population of total size at time t given by $N(t)$ in which individuals are recruited at a total recruitment rate Λ and die at a natural death rate μ. The number of healthy individuals who are susceptible to the disease at time t is denoted by $S(t)$. Healthy individuals can get infected by strain one at a transmission rate β_1 and enter the class of individuals infected and infectious with strain one. This class is of total size $I_1(t)$. Independently, healthy individuals can get infected by strain two at a transmission rate β_2 and enter the class of individuals infected and infectious with strain two whose total size is given by $I_2(t)$. Individuals, infected with strain one can get infected with strain two from those infected with strain two only at a rate δ_1 while individuals infected with strain two can get infected with strain one from those infected with strain one only at a rate δ_2 — those individuals become jointly infected with both strains and enter the class $J(t)$. Jointly infected individuals infect susceptibles with strain one at a rate γ_1 and those in the I_1 class with strain two — at a rate η_1. Similarly, jointly infected individuals infect susceptibles with strain two at a rate γ_2 and those in the I_2 class with strain one — at a rate η_2. Infected with strain one recover at a rate α_1, those infected with strain two — at a rate α_2 and those jointly infected recover approximately at the same time from both strains at a rate ν. All recovered individuals make up the class $R(t)$. Finally, susceptible individuals are vaccinated at a vaccination rate ψ and enter the class of vaccinated individuals, $V(t)$. We note that all vaccinated individuals are fully protected against all strains in the system, that is, we assume "perfect" vaccination. The model takes the form:

$$
\begin{aligned}
S' &= \Lambda - \beta_1 \frac{SI_1}{N} - \beta_2 \frac{SI_2}{N} - (\gamma_1 + \gamma_2)\frac{SJ}{N} - (\mu + \psi)S \\
I_1' &= \beta_1 \frac{SI_1}{N} + \gamma_1 \frac{SJ}{N} - (\mu + \alpha_1)I_1 - \delta_1 \frac{I_1 I_2}{N} - \eta_1 \frac{I_1 J}{N} \\
I_2' &= \beta_2 \frac{SI_2}{N} + \gamma_2 \frac{SJ}{N} - (\mu + \alpha_2)I_2 - \delta_2 \frac{I_1 I_2}{N} - \eta_2 \frac{I_2 J}{N} \\
J' &= \delta_1 \frac{I_1 I_2}{N} + \eta_1 \frac{I_1 J}{N} + \delta_2 \frac{I_1 I_2}{N} + \eta_2 \frac{I_2 J}{N} - (\mu + \nu)J \\
R' &= \alpha_1 I_1 + \alpha_2 I_2 + \nu J - \mu R \\
V' &= \psi S - \mu V
\end{aligned}
\tag{4.1}
$$

The existence of equilibria depends on the reproduction numbers of the two strains which are symmetrical:

$$
\mathcal{R}_1(\psi) = \frac{\beta_1 \mu}{(\mu + \psi)(\mu + \alpha_1)} \qquad \mathcal{R}_2(\psi) = \frac{\beta_2 \mu}{(\mu + \psi)(\mu + \alpha_2)} \tag{4.2}
$$

We note again that both reproduction numbers are decreasing functions of the vaccination rate ψ. In this case both can be decreased to zero by vaccination. The system (4.1) has the disease-free equilibrium

$$
\mathcal{E}_0 = \left(\frac{\mu}{\mu + \psi}, 0, 0, 0, 0, \frac{\psi}{\mu + \psi} \right),
$$

where each equilibrium consists of the proportion of susceptible, proportion of infected with the first strain, proportion infected with the second strain, proportion of jointly infected, proportion recovered, and proportion vaccinated individuals: $\mathcal{E} = (s, i_1, i_2, j, r, v)$. The system has two dominance equilibria - one for each strain. The dominance equilibrium of strain one exists if and only if $\mathcal{R}_1(\psi) > 1$ and is given by:

$$
\mathcal{E}_1 = \left(\frac{1}{\mathcal{R}_1}, \frac{\mu}{\mu + \alpha_1}\left(1 - \frac{1}{\mathcal{R}_1(\psi)}\right), 0, 0, \frac{\alpha_1}{\mu + \alpha_1}\left(1 - \frac{1}{\mathcal{R}_1(\psi)}\right), \frac{\psi}{\mu \mathcal{R}_1} \right)
$$

Similarly, dominance equilibrium of strain two exists if and only if $\mathcal{R}_2(\psi) > 1$ and is given by:

$$
\mathcal{E}_2 = \left(\frac{1}{\mathcal{R}_2}, \frac{\mu}{\mu + \alpha_2}\left(1 - \frac{1}{\mathcal{R}_2(\psi)}\right), 0, 0, \frac{\alpha_2}{\mu + \alpha_2}\left(1 - \frac{1}{\mathcal{R}_2(\psi)}\right), \frac{\psi}{\mu \mathcal{R}_2} \right)
$$

Denote by $q_i = \frac{\eta_i}{\mu+\nu}$ and $r_i = \frac{\gamma_i}{\mu+\nu}$ for $i = 1, 2$. The invasion reproduction number of the first strain at the equilibrium of the second is given by

$$
\hat{\mathcal{R}}_1(\psi) = \frac{\beta_1 s + r_1 s(\delta_1 + \delta_2)i_2 + q_2(\mu + \alpha_1 + \delta_1 i_2)i_2}{\mu + \alpha_1 + \delta_1 i_2 + q_2 i_2 \beta_1 s} \tag{4.3}
$$

where s and i_2 have the corresponding values from $\mathcal{E}_2$. Analogously, the reproduction number of the second strain at the equilibrium of the first is given by

$$\hat{\mathcal{R}}_2(\psi) = \frac{\beta_2 s + r_2 s(\delta_1 + \delta_2)i_1 + q_1(\mu + \alpha_2 + \delta_2 i_1)i_1}{\mu + \alpha_2 + \delta_2 i_1 + q_1 i_1 \beta_2 s} \tag{4.4}$$

where s and i_1 have the corresponding values from $\mathcal{E}_1$. Both invasion reproduction numbers depend on ψ only through i_1 and i_2. Both i_1 and i_2 are decreasing functions of ψ but the dependence of the invasion reproduction numbers on i_1 and i_2 may be non-monotone. In particular, the dependence of the invasion reproduction numbers on i_1 and i_2 may be monotonely decreasing, monotonely increasing or first monotonely decreasing and then monotonely increasing. That translates to exactly opposite dependence of the invasion numbers on ψ. First, we will assume without loss of generality that

$$\mathcal{R}_1 < \mathcal{R}_2.$$

This, in particular means that in the absence of the trade-off mechanism (coinfection in this case) strain two will dominate in the population both in absence and in presence of vaccination of any level. Second, since the intrinsic reproduction number of the second strain is larger, to break the symmetry of the strains we make the following assumptions that strengthen strain one and weaken strain two:

Assumption 4.1.

1. *Suppose that strain one can coinfect individuals infected with strain two but strain two cannot coinfect individuals infected with strain one. That, in particular, means that we are assuming:*

$$\delta_1 = 0, \quad \eta_1 = 0 \ (q_1 = 0).$$

2. *Suppose that jointly infected individuals, that is those in class J cannot infect with strain two, that is,*

$$\gamma_2 = 0 \ (r_2 = 0).$$

Under these assumptions the invasion reproduction numbers become:

$$\hat{\mathcal{R}}_1(\psi) = \frac{\beta_1 s + r_1 s \delta_2 i_2 + q_2(\mu + \alpha_1)i_2}{\mu + \alpha_1 + q_2 i_2 \beta_1 s} \tag{4.5}$$

The reproduction number of the second strain at the equilibrium of the first becomes:

$$\hat{\mathcal{R}}_2(\psi) = \frac{\beta_2 s}{\mu + \alpha_2 + \delta_2 i_1} \tag{4.6}$$

The invasion reproduction number of the second strain $\hat{\mathcal{R}}_2(\psi)$ is now a decreasing function of i_1, and consequently, an increasing function of ψ. On the other hand, the derivative of $\hat{\mathcal{R}}_1(\psi)$ with respect to i_2 is

$$\frac{\partial \hat{\mathcal{R}}_1(\psi)}{\partial i_2} = \frac{r_1\delta_2(\mu + \alpha_1)s + q_2(\mu + \alpha_1)^2 \left[1 - \left(\frac{\mathcal{R}_1}{\mathcal{R}_2}\right)^2\right]}{(\mu + \alpha_1 + q_2 i_2 \beta_1 s)^2} > 0$$

Consequently, $\hat{\mathcal{R}}_1(\psi)$ is an increasing function of i_2 and therefore a decreasing function of ψ. This implies that vaccination has reciprocal effect of the invasion reproduction numbers. In particular, it decreases the invasion capabilities of the first strain and increases the invasion capabilities of the second strain. Thus, if coinfection allows the first strain to dominate in the population when no vaccination is present, that is, $\hat{\mathcal{R}}_1(0) > 1$ while $\hat{\mathcal{R}}_2(0) < 1$, increasing vaccination levels may lead to the fact that at some vaccination level $\hat{\psi}$ we have $\hat{\mathcal{R}}_1(\hat{\psi}) < 1$ and $\hat{\mathcal{R}}_2(\hat{\psi}) > 1$, that is strain two dominates in the population. Replacement of strain one with strain two has occurred. We illustrate that in Figure 2. We note that in Figure 2 we have the case when $\mathcal{R}_2 > \mathcal{R}_1$, and consequently, in the absence of coinfection strain two will be dominating (with or without vaccination). In the absence of vaccination $\psi = 0$, the trade-off mechanism is strong enough to allow for the strain with the lower reproduction number to persist while the one with the larger reproduction number is eliminated. That is a result of the fact that the first strain can invade the equilibrium of the second $\hat{\mathcal{R}}_1(0) = 1.44$ while the second cannot invade the equilibrium of the first $\hat{\mathcal{R}}_2(0) = 0.9091$. This outcome is a result of our Assumptions 4.1 that strengthen the first strain in its interaction with the second. Furthermore, vaccination decreases the impact of the trade-off mechanism and restores the strain with the larger intrinsic reproduction number to dominate in the population.

This example raises several questions: Is it absolutely necessary that the trade-off mechanism is strong enough to allow the strain with the lower reproduction number to dominate in the population. Super-infection [25] and coinfection [21] are two such mechanism which are known to generate this effect but not all trade-off mechanisms can be readily associated with it. In particular, we have previously observed that cross-immunity as a trade-off mechanism always leads to dominance of the strain with the largest reproduction number [26]. Before we address the question whether cross-immunity may trigger strain replacement, we will investigate whether strain replacement may be exhibited under a different scenario. In particular, assume it is possible that the trade-off mechanism is weak and does

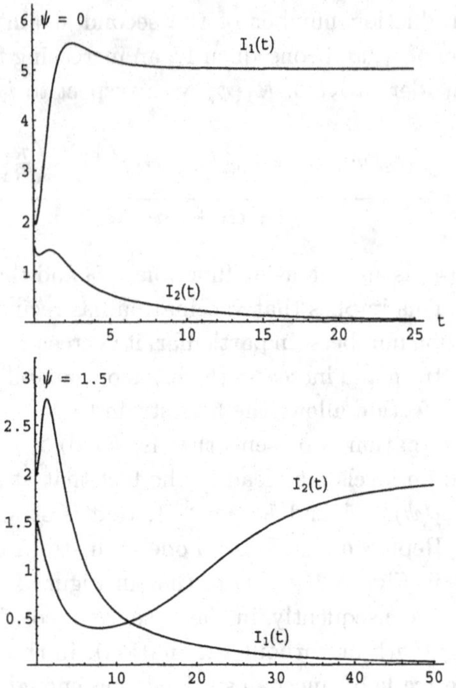

Figure 2. The figure shows that strain replacement occurs in the model with coinfection (4.1). The left figure shows that strain one dominates while strain two is eliminated when there is no vaccination $\psi = 0$. The right figure shows that strain two dominates while strain one is eliminated when vaccination is at level $\psi = 1.5$. The remaining parameters in this figure are chosen as follows: $\beta_1 = 6$, $\beta_2 = 5$, $\mu = 0.5$, $\gamma_1 = 8$, $\gamma_2 = 0$, $\delta_1 = 0$, $\delta_2 = 1.5$, $\alpha_1 = 1$, $\alpha_2 = 0.5$, $\nu = 0.5$, $\eta_1 = 0$, $\eta_2 = 0$, $\Lambda = 10$. These give $\mathcal{R}_1 = 4$ and $\mathcal{R}_2 = 5$.

not lead to dominance of the strain with the smaller intrinsic reproduction number. Is it possible that a "perfect" vaccination works in such a way that it strengthens this weak trade-off mechanism so that at some higher vaccination levels this mechanism allows for the strain with the lower intrinsic reproduction number to dominate? In the case of coinfection we could answer this question negatively only for a special case.

Proposition 4.2. *Assume that jointly infected individuals cannot coinfect already infected individuals, that is* $\eta_1 = \eta_2 = 0$. *Assume that in the absence of vaccination* ($\psi = 0$) *the strain with the larger reproduction number persists while the other is eliminated. Then strain replacement cannot occur in model (4.1).*

Proof. Assume without loss of generality that $\mathcal{R}_1 > \mathcal{R}_2$, and that $\hat{\mathcal{R}}_1(0) > 1$ while $\hat{\mathcal{R}}_2(0) < 1$. The invasion reproduction number of the first strain at the equilibrium of the second in this case becomes

$$\hat{\mathcal{R}}_1(\psi) = \frac{\beta_1 s + r_1 s(\delta_1 + \delta_2) i_2}{\mu + \alpha_1 + \delta_1 i_2} \tag{4.7}$$

where we recall that s and i_2 have the corresponding values from $\mathcal{E}_2$. Analogously, the invasion reproduction number of the second strain at the equilibrium of the first is given by

$$\hat{\mathcal{R}}_2(\psi) = \frac{\beta_2 s + r_2 s(\delta_1 + \delta_2) i_1}{\mu + \alpha_2 + \delta_2 i_1} \tag{4.8}$$

where we recall s and i_1 have the corresponding values from $\mathcal{E}_1$. The derivatives of the two invasion reproduction numbers with respect to i_2 and i_1 respectively, are given by

$$\begin{aligned}\frac{\partial \hat{\mathcal{R}}_1(\psi)}{\partial i_2} &= \frac{r_1 s(\delta_1 + \delta_2)(\mu + \alpha_1) - \delta_1 \beta_1 s}{(\mu + \alpha_1 + \delta_1 i_2)^2} \\ &= \frac{s(\mu + \alpha_1)[r_1(\delta_1 + \delta_2) - \delta_1 \mathcal{R}_1]}{(\mu + \alpha_1 + \delta_1 i_2)^2}\end{aligned}$$

Analogously,

$$\begin{aligned}\frac{\partial \hat{\mathcal{R}}_2(\psi)}{\partial i_1} &= \frac{r_2 s(\delta_1 + \delta_2)(\mu + \alpha_2) - \delta_2 \beta_2 s}{(\mu + \alpha_2 + \delta_2 i_1)^2} \\ &= \frac{s(\mu + \alpha_2)[r_2(\delta_1 + \delta_2) - \delta_2 \mathcal{R}_2]}{(\mu + \alpha_2 + \delta_2 i_1)^2}\end{aligned}$$

Assume that replacement occurs. That implies that there exists a vaccination level ψ^* such that $\hat{\mathcal{R}}_1(\psi^*) = 1$ becoming from larger than one to smaller than one. This equality can be rewritten to give

$$(\mu + \alpha_1)\left[\frac{\mathcal{R}_1}{\mathcal{R}_2} - 1\right] = i_2[\delta_1 - r_1 s(\delta_1 + \delta_2)]$$

From our assumption that $\mathcal{R}_1 > \mathcal{R}_2$ it follows that the left-hand side of this equality is positive. Since $i_2 > 0$, that means that we must have

$$\delta_1 > r_1 s(\delta_1 + \delta_2)$$

which implies that

$$\delta_1 \mathcal{R}_2 > r_1(\delta_1 + \delta_2)$$

or that $\frac{\partial \hat{\mathcal{R}}_1(\psi)}{\partial i_2} < 0$ leading to the fact that $\hat{\mathcal{R}}_1(\psi)$ is an increasing function of ψ starting from a value above one and cannot be equal to one for any ψ. That is a contradiction. Consequently, strain replacement does not occur. □

5. Cross-Immunity and Strain Replacement

Cross-immunity is the phenomenon where an infection with one strain gives partial protection to infection with other strains. Cross-immunity has been primarily investigated in connection with influenza [1, 5, 6] and dengue [7, 10, 17]. It is well known that it can cause coexistence of the strains. Imperfect vaccination can lead to strain replacement in conjunction with cross-immunity only if the strains provide sufficient level of protection against each other, that is, when the competition among them is high [8, 13]. Here we investigate the possibility that cross-immunity as a trade-off mechanism may cause replacement in the context of "perfect" vaccination. From the analysis in [26] we know that the boundaries of the coexistence region do not cross the bisector $\mathcal{R}_1 = \mathcal{R}_2$. That, in particular, implies that only the strain with the higher reproduction number dominates. Here, we use a cross-immunity model similar to the one in [5, 6] and somewhat simpler than the one in [26]. We use that model to show that "perfect" vaccination combined with cross-immunity in its simplest form cannot lead to strain replacement. This suggests that not all trade-off mechanisms necessarily induce replacement effect.

We consider again a host population of total size at time t given by $N(t)$ with recruitment Λ and natural death rate μ. The class of susceptible to the disease individuals at time t is denoted by $S(t)$. Healthy individuals who previously were never infected by any of the strains can get infected by strain one at a rate β_1 and enter the class of individuals infected and infectious with strain one $I_1(t)$. Those recover at a rate γ_1 and enter the recovered class $R_1(t)$. Recovered individuals in class $R_1(t)$ cannot get infected with strain one any more but they can get infected by strain two at a somewhat reduced rate $\beta_2\sigma$, where σ is the cross-immunity coefficient, and move to the class of individuals currently infected and infectious with strain two who were previously infected with train one $J_2(t)$. Individuals who recover from $J_2(t)$ enter the class of fully immune individuals $W(t)$ at a rate γ_2. This same process can be symmetrically undergone through a first infection with strain two giving rise to the analogous classes $I_2(t)$, $R_2(t)$, and $J_1(t)$. Finally, susceptible individuals are vaccinated at a vaccination rate ψ and enter the class of vaccinated individuals, $V(t)$. We note that all vaccinated individuals are fully protected against all strains in the system, that is, we again assume "perfect" vaccination. We obtain the following model:

$$\begin{aligned}
S' &= \Lambda - \beta_1 \frac{S(I_1+J_1)}{N} - \beta_2 \frac{S(I_2+J_2)}{N} - (\mu+\psi)S \\
I_1' &= \beta_1 \frac{S(I_1+J_1)}{N} - (\mu+\gamma_1)I_1 \\
R_1' &= \gamma_1 I_1 - \beta_2 \sigma R_1 \frac{(I_2+J_2)}{N} - \mu R_1 \\
J_1' &= \beta_1 \sigma R_2 \frac{(I_1+J_1)}{N} - (\mu+\gamma_1)J_1 \\
I_2' &= \beta_2 \frac{S(I_2+J_2)}{N} - (\mu+\gamma_2)I_2 \\
R_2' &= \gamma_2 I_2 - \beta_1 \sigma R_2 \frac{(I_1+J_1)}{N} - \mu R_2 \\
J_2' &= \beta_2 \sigma R_1 \frac{(I_2+J_2)}{N} - (\mu+\gamma_2)J_2 \\
W' &= \gamma_1 J_1 + \gamma_2 J_2 - \mu W \\
V' &= \psi S - \mu V
\end{aligned} \tag{5.1}$$

The reproduction numbers of the two strains are the same as in the coinfection case and are symmetrical:

$$\mathcal{R}_1(\psi) = \frac{\beta_1 \mu}{(\mu+\psi)(\mu+\alpha_1)} \qquad \mathcal{R}_2(\psi) = \frac{\beta_2 \mu}{(\mu+\psi)(\mu+\alpha_2)} \tag{5.2}$$

The system (5.1) has the disease-free equilibrium

$$\mathcal{E}_0 = \left(\frac{\mu}{\mu+\psi}, 0, 0, 0, 0, 0, 0, 0, \frac{\psi}{\mu+\psi} \right),$$

where each equilibrium consists of the following proportions:

$\mathcal{E} = (s, i_1, r_1, j_1, i_2, r_2, j_2, w, v)$. The system has two dominance equilibria- one for each strain. The dominance equilibrium of strain one exists if and only if $\mathcal{R}_1(\psi) > 1$ and is given by:

$$\mathcal{E}_1 = \left(\frac{1}{\mathcal{R}_1}, \frac{\mu}{\mu+\gamma_1}\left(1 - \frac{1}{\mathcal{R}_1(\psi)}\right), \frac{\gamma_1}{\mu+\alpha_1}\left(1 - \frac{1}{\mathcal{R}_1(\psi)}\right), 0, 0, 0, 0, 0, \frac{\psi}{\mu \mathcal{R}_1} \right)$$

Similarly, the dominance equilibrium of strain two exists if and only if $\mathcal{R}_2(\psi) > 1$ and is given by:

$$\mathcal{E}_2 = \left(\frac{1}{\mathcal{R}_2}, 0, 0, 0, \frac{\mu}{\mu+\gamma_2}\left(1 - \frac{1}{\mathcal{R}_2(\psi)}\right), \frac{\gamma_2}{\mu+\gamma_2}\left(1 - \frac{1}{\mathcal{R}_2(\psi)}\right), 0, 0, \frac{\psi}{\mu \mathcal{R}_2} \right)$$

The invasion reproduction number of the first strain at the equilibrium of the second strain is given by

$$\hat{\mathcal{R}}_1(\psi) = \frac{\mathcal{R}_1}{\mathcal{R}_2} + \frac{\mathcal{R}_1\sigma\gamma_2}{\mu+\gamma_2}\left(1 - \frac{1}{\mathcal{R}_2(\psi)}\right) \tag{5.3}$$

The invasion reproduction number of the second strain at the equilibrium of the first strain is given by

$$\hat{\mathcal{R}}_2(\psi) = \frac{\mathcal{R}_2}{\mathcal{R}_1} + \frac{\mathcal{R}_2\sigma\gamma_1}{\mu+\gamma_1}\left(1 - \frac{1}{\mathcal{R}_1(\psi)}\right) \tag{5.4}$$

Clearly both invasion reproduction numbers are decreasing functions of the vaccination rate and thus vaccination does not have a reciprocal effect on the invasion capabilities of the strains. In fact, it decreases both. Consequently, strain replacement in the strong form that we are considering in this article — the dominance of one strain is replaced by dominance of the other — does not occur. It is possible that *weak replacement* in the form of Scenario 1 or Scenario 2 may occur. For instance, it is possible that one of the strains has a much higher prevalence but the other has a much lower prevalence while the two strains coexist and with increased vaccination levels the strain with the higher prevalence gets eliminated first and the other strain dominates.

In this context, strain replacement in the stronger sense considered here does not occur in the presence of another trade-off mechanism — mutation — defined as one strain changing its genetic characteristic to become another (and the host infected with it becomes a host host infected with the new strain) [2]. It is well known that mutation leads to coexistence, but the newly obtained mutant strain cannot exist by itself, that is, it cannot be a dominant strain [9]. Therefore, in the case of mutation strain replacement in the strong sense considered here does not occur.

6. Discussion

In this article we investigate the role of vaccination in strain replacement. We understand the replacement effect in a strong sense: we assume that one of the strains dominates in the absence of vaccination, while in the presence of vaccination — the other strain dominates. We call this vaccine induced replacement effect since vaccination is necessary to bring about the other strain.

The replacement effect has been thoroughly investigated in the literature — there are both plenty of empirical evidence and theoretical studies.

Mathematical models have been used to investigate how and why it occurs. It has been suggested that the replacement effect is a result of the differential effectiveness of the vaccine.

In this article we add to an already existing theoretical evidence that differential effectiveness of the vaccine may not be necessary for a vaccine induced replacement effect to occur. We suggest a new mechanism that may explain why strain replacement under vaccination may occur. In particular, we suggest that vaccination leads to exchange in the dominance of strains because it has a reciprocal effect on the fitness of the strains, that is because it decreases the fitness of the strain dominating in the absence of vaccination, and it increases the fitness of the strain dominating in the presence of vaccination. We define the fitness of the strain as its capability to invade the equilibrium of the other strain, that is, its reproduction number when the other strain is at equilibrium, given by the invasion reproduction number.

Furthermore, exchange of dominance of the strains through vaccination appears to be possible only if the strains have the ability to coexist. Therefore, exchange of dominance is strongly connected to the action of some trade-off mechanism. In fact, in all known theoretical cases that detect the phenomenon, stable coexistence of the strains is also possible as well as competitive exclusion. To support that claim we establish that differential effectiveness is a trade-off mechanism itself by showing that in the absence of other known trade-off mechanisms equally effective vaccines lead to competitive exclusion of the strain with the lower reproduction number. On the other hand, differential effectiveness of the vaccine leads to (locally stable) coexistence. From this perspective it is hardly surprising that some other trade-off mechanisms in combination with equally effective, and even "perfect", vaccines can also lead to strain replacement. We observed it an epidemic model with perfect vaccination and super-infection as a trade-off mechanism. We show that confection as a trade-off mechanism combined with "perfect" vaccination can also lead to strain replacement in the strong sense we consider here. On the other hand strain replacement in the strong sense does not occur with perfect vaccination in the presence of several well-known trade-off mechanisms, such as cross-immunity and mutation. Perhaps, it may occur in some weakened sense where the strains coexist but exchange their position as the most prevalent strain. We did not explore that further because such exploration requires knowledge of the mechanism that governs higher prevalence during coexistence.

So which trade-off mechanisms can lead to strain replacement in the presence of equally effective vaccines and which cannot? We surmise that the trade-off mechanism should be strong enough to be capable to allow a strain with a lower intrinsic reproduction number to dominate in the population. Both super-infection and coinfection are known to lead to extinction of the strain with the maximal reproduction number even in the absence of vaccination. On the other hand, there is no evidence that cross-immunity may cause such effect.

"Perfect" vaccination coupled with either super-infection or coinfection seems to lead to strain replacement through exactly the same sequence of steps: The strain with the largest intrinsic reproduction number will dominate in the absence of vaccination and the trade-off mechanism. However, in the presence of the trade-off mechanism but in the absence of vaccination — the strain with the lower intrinsic reproduction number dominates and the strain with the larger reproduction is eliminated. Vaccination weakens the effect of the trade-off mechanism and at some vaccination level the dominance of the strain with the larger intrinsic reproduction number is restored. Thus we see replacement of the dominance of the strain with the lower intrinsic reproduction number with the strain with the larger intrinsic reproduction number. We have not been able to show that "perfect" vaccination can lead to replacement of the strain with the larger intrinsic reproduction number by a strain with a lower intrinsic reproduction number. We have been able to rule out this possibility for a special case of coinfection, however, ruling it out for the more general cases remains an open problem.

This is where the replacement effect that occurs with differential effectiveness of the vaccine differs significantly compared to the one that occurs with "perfect" vaccines. Aside the fact that most vaccines are indeed unequally effective against different strains — differential effectiveness leads to the replacement of the strain with the higher reproduction number which dominates in the absence of vaccination (when competitive exclusion is the only outcome) with the strain with a lower reproduction number (when unequally effective vaccine acts as a trade-off mechanism). Another marked difference is that even if "perfect" vaccines can cause replacement, that can only happen for a certain range of vaccination levels. If the vaccination level is sufficiently high — both strains will be eliminated from the population. That may not be the case with differentially effective vaccines. If the reproduction number of the strain, not primarily targeted by the vaccine, cannot be reduced below one, no matter how high the vaccination levels, that strain will persist, even if we successfully vaccinate all individuals in the population.

Acknowledgments

MM was visiting Department of Zoology, UF when this paper was written. The paper benefited significantly from interaction with faculty members and, particularly, from discussions with B. Bolker. MM was supported by NSF grant DMS-0408230.

References

[1] V. Andreasen, J. Lin, and S. Levin. The dynamics of cocirculating influenza strains conferring partial cross-immunity, *J. Math. Biol.*, **35** (1997) 825–842.

[2] S. Bonhoeffer and M. Nowak. Mutation and the evolution of virulence, *Proc. Royal Soc. London B*, **258** (1994) 133–140.

[3] H.-J. Bremermann and H. R. Thieme. A competitive exclusion principle for pathogen virulence, *J. Math. Biol.*, **27** (1989) 179–190.

[4] N. Cardeñosa, A. Domínguez, A. Martínez, J. Alvarez, H. Pañella, P. Godoy, S. Minguell, N. Camps, and J. A. Vázquez. Meningococcal disease in Catalonia 1 year after mass vaccination campaign with meningococcal group C polysaccharide vaccine, *Infection*, **31**(6) (2003) p. 392–397.

[5] C. Castillo-Chavez, H. Hethcote, V. Andreasen, S. Levin, and W. M. Liu. Epidemiological models with age structure, proportionate mixing and cross-immunity, *J. Math. Biol.*, **27** (1989) 159–165.

[6] C. Castillo-Chavez, H. Hethcote, V. Andreasen, S. Levin, and W. M. Liu. Cross-immunity in the dynamics of homogeneous and heterogeneous populations, *Mathematical Ecology* (Trieste, 1986), World Sci. Publishing, Teaneck, NJ, (1988) 303–316.

[7] L. Esteva and C. Vargas. Coexistence of different serotypes of dengue virus, *J. Math. Biol.*, **46** (2003) 31–47.

[8] E. H. Elbasha and A. P. Galvani. Vaccination against multiple HPV types, *Math. Biosci.* **197** (2005) 88–117.

[9] Z. Feng, M. Iannelli, and F. Milner. A two-strain tuberculosis model with age of infection, *SIAM J. Appl. Math.* **62** (2002) 1634–1656.

[10] N. Ferguson, R. Anderson, and S. Gupta. The effect of atibody-dependent enhancement on the transmission dynamics and persistence of multiple-strain pathogens, *Proc. Natl. Acad. Sci. USA*, **96** (1999) 790–794.

[11] S. Gandon, M. J. Mackinnon, S. Nee, and A. F. Read. Imperfect vaccines and the evolution of pathogen virulence, *Nature* **414** (2001) 751–756.

[12] B. T. Grenfell, O. G. Pybus, J. R. Gog, J. L. N. Wood, J. M. Daly, J. A. Mumford, and E. C. Holmes. Unifying the epidemiological and evolutionary dynamics of pathogens, *Science*, **303** (2004) 327–332.

[13] S. Gupta, N. M. Ferguson, and R. M. Anderson. Vaccination and the population structure of antigenically diverse pathogens that exchange genetic material, *Proc. R. Soc. Lond.*, **B264** (1997) 1435–1443.

[14] K. P. Hadeler and Castillo-C. Chavez. A core group model for disease transmission, *Math. Biosci.*, **128** (1995) 41–55.

[15] S. S. Huang, R. Platt, Rifas-S. L. Shiman, S. I. Pelton, D. Goldman, and J. A. Finkelstein. Post-PCV7 changes in colonzing pnemococcal serotypes in 16 Massachusetts communities, 2001 and 2004, *Pediatrics* **116**(3) (2005) p. 408–413.
[16] M. Iannelli, M. Martcheva, and X.-Z. Li. Strain replacement in an epidemic model with super-infection and perfect vaccination, *Math. Biosci.*, **195** (2005) p. 23–46.
[17] I. Kawaguchi, A. Sasaki, and M. Boots. Why are dengue virus serotypes so distantly related? Enhancement and limiting serotype similarity between dengue virus strains, *Proc. Royal Soc. London, B*, **270**(1530) (2003) 2241–2247.
[18] J. Li, Y. Zhou, Zh. Ma, and J. M. Hyman. Epidemiological models for mutating pathogens, *SIAM J. Appl. Math.*, **65** (2004) 1–23.
[19] M. Lipsitch. Vaccination against colonizing bacteria with multiple serotypes, *Proc. Natl. Acad. Sci. USA*, **94** (1997) p. 6571–6576.
[20] M. Lipsitch. Bacterial vaccines and serotype replacement: lessons from *Haemophilus influenzae* and prospects for *Streptococcus pneumoniae*, *Emerg. Inf. Dis.*, **5**(3) (1999) p. 336–345.
[21] M. Martcheva and S. S. Pilyugin. The role of coinfection in multi-disease dynamics, *SIAM J. Appl. Math.*, (to appear).
[22] R. May and M. Nowak. Coinfection and the evolution of parasite virulence, *Proc. Royal Soc. London B*, **261** (1995) 209–215.
[23] A. McLean. Development and use of vaccines against evolving pathogens: vaccine design, in *Evolution in Health and Disease*, (S. C. Stearns, ed.), 138–151, Oxford University Press, Oxford, 1999.
[24] J. D. Miller. The Replacement effect, *The Scientist*, May 23, 2003.
[25] M. Nowak and R. May. Superinfection and the evolution of parasite virulence, *Proc. Royal Soc. London B*, **255** (1994) 81–89.
[26] M. Nuño, Z. Feng, M. Martcheva, and C. Castillo-Chavez. Dynamics of two-strain influenza with isolation and partial cross-immunity, *SIAM J. Appl. Math.*, **65**(3) (2005) 964–982.
[27] T. C. Porco and S. M. Blower. Designing HIV vaccination policies: subtypes and cross-immunity, *Interfaces* **28**(3) (1998) p. 167–190.
[28] T. C. Porco and S. M. Blower. HIV vaccines: The Effect of the mode of action on the coexistence of HIV subtypes, *Math. Pop. Studies*, **8**(2) (2000) p. 205–229.

SERIES ON KNOTS AND EVERYTHING

Editor-in-charge: Louis H. Kauffman *(Univ. of Illinois, Chicago)*

The Series on Knots and Everything: is a book series polarized around the theory of knots. Volume 1 in the series is Louis H Kauffman's Knots and Physics.

One purpose of this series is to continue the exploration of many of the themes indicated in Volume 1. These themes reach out beyond knot theory into physics, mathematics, logic, linguistics, philosophy, biology and practical experience. All of these outreaches have relations with knot theory when knot theory is regarded as a pivot or meeting place for apparently separate ideas. Knots act as such a pivotal place. We do not fully understand why this is so. The series represents stages in the exploration of this nexus.

Details of the titles in this series to date give a picture of the enterprise.

Published:

Vol. 1: Knots and Physics (3rd Edition)
by L. H. Kauffman

Vol. 2: How Surfaces Intersect in Space — An Introduction to Topology (2nd Edition)
by J. S. Carter

Vol. 3: Quantum Topology
edited by L. H. Kauffman & R. A. Baadhio

Vol. 4: Gauge Fields, Knots and Gravity
by J. Baez & J. P. Muniain

Vol. 5: Gems, Computers and Attractors for 3-Manifolds
by S. Lins

Vol. 6: Knots and Applications
edited by L. H. Kauffman

Vol. 7: Random Knotting and Linking
edited by K. C. Millett & D. W. Sumners

Vol. 8: Symmetric Bends: How to Join Two Lengths of Cord
by R. E. Miles

Vol. 9: Combinatorial Physics
by T. Bastin & C. W. Kilmister

Vol. 10: Nonstandard Logics and Nonstandard Metrics in Physics
by W. M. Honig

Vol. 11: History and Science of Knots
edited by J. C. Turner & P. van de Griend

Vol. 12: Relativistic Reality: A Modern View
edited by J. D. Edmonds, Jr.

Vol. 13: Entropic Spacetime Theory
by J. Armel

Vol. 14: Diamond — A Paradox Logic
by N. S. Hellerstein

Vol. 15: Lectures at KNOTS '96
by S. Suzuki

Vol. 16: Delta — A Paradox Logic
by N. S. Hellerstein

Vol. 17: Hypercomplex Iterations — Distance Estimation and Higher Dimensional Fractals
by Y. Dang, L. H. Kauffman & D. Sandin

Vol. 19: Ideal Knots
by A. Stasiak, V. Katritch & L. H. Kauffman

Vol. 20: The Mystery of Knots — Computer Programming for Knot Tabulation
by C. N. Aneziris

Vol. 24: Knots in HELLAS '98 — Proceedings of the International Conference on Knot Theory and Its Ramifications
edited by C. McA Gordon, V. F. R. Jones, L. Kauffman, S. Lambropoulou & J. H. Przytycki

Vol. 25: Connections — The Geometric Bridge between Art and Science (2nd Edition)
by J. Kappraff

Vol. 26: Functorial Knot Theory — Categories of Tangles, Coherence, Categorical Deformations, and Topological Invariants
by David N. Yetter

Vol. 27: Bit-String Physics: A Finite and Discrete Approach to Natural Philosophy
by H. Pierre Noyes; edited by J. C. van den Berg

Vol. 28: Beyond Measure: A Guided Tour Through Nature, Myth, and Number
by J. Kappraff

Vol. 29: Quantum Invariants — A Study of Knots, 3-Manifolds, and Their Sets
by T. Ohtsuki

Vol. 30: Symmetry, Ornament and Modularity
by S. V. Jablan

Vol. 31: Mindsteps to the Cosmos
by G. S. Hawkins

Vol. 32: Algebraic Invariants of Links
by J. A. Hillman

Vol. 33: Energy of Knots and Conformal Geometry
by J. O'Hara

Vol. 34: Woods Hole Mathematics — Perspectives in Mathematics and Physics
edited by N. Tongring & R. C. Penner